T0214029

# Hardware-Aware Probabilistic Machine Learning Models

Laura Isabel Galindez Olascoaga • Wannes Meert
Marian Verhelst

# Hardware-Aware Probabilistic Machine Learning Models

## Learning, Inference and Use Cases

 Springer

Laura Isabel Galindez Olascoaga
Electrical Engineering
KU Leuven
Leuven, Belgium

Wannes Meert
Computer Science
KU Leuven
Leuven, Belgium

Marian Verhelst
Electrical Engineering
KU Leuven
Leuven, Belgium

ISBN 978-3-030-74044-3        ISBN 978-3-030-74042-9   (eBook)
https://doi.org/10.1007/978-3-030-74042-9

This Springer imprint is published by the registered company Springer Nature Switzerland AG
The registered company address is: Gewerbestrasse 11, 6330 Cham, Switzerland

# Preface

Throughout the last decade, cloud-computing paradigms have proven successful at exploiting the opportunities and addressing the challenges of the ever-increasing number of electronic devices populating our environment. However, the privacy, latency, and efficiency concerns of this approach have motivated users and developers to seek computing paradigms that remain closer to the data-collecting devices, otherwise known as edge devices.

One of the main challenges to overcome in the quest towards edge-computing is the significant energy and computational bandwidth constraints of the involved devices, which tend to be battery operated and portable, especially when devices have to execute complex algorithms relying on large amounts of high-quality sensory data, such as in the case with machine learning applications. Furthermore, smart portable devices are prone to dynamically changing environmental conditions and noise, and the sensory data they rely on is inherently uncertain.

Probabilistic models constitute a suitable approach to address some of these challenges, since they can represent the inherent uncertainty of edge applications, and are robust to noise and missing data. However, their implementation in resource-constrained edge devices has not been studied extensively, unlike other machine learning paradigms, such as neural networks, which have already seen tremendous progress in this research field.

This book proposes to endow probabilistic models with *hardware-awareness*, in an attempt to enable their efficient implementation in resource-constrained edge devices. These models can represent scalable properties of the devices that host them, such as the quality with which their sensors operate, or the complexity of the inference algorithm they ought to perform.

The strategies proposed herewith can exploit these models to evaluate the impact that a specific device configuration may have on the machine learning task's resource consumption and performance, with the overarching goal of balancing the two optimally. The proposed models can also consider various sub-systems and their properties holistically, bringing about resource-saving opportunities that other resource-aware approaches fail to uncover. Specifically, this book focuses on augmenting Bayesian networks and probabilistic circuits with hardware-awareness.

These models' performance is empirically evaluated in several use cases that consider different types of systems and various scalable properties. The range of use cases proves to benefit from this thesis's contributions, with the potential of attaining significant resource-saving opportunities with minimal accuracy losses at application time.

Leuven, Belgium                                                    Laura Isabel Galindez Olascoaga
November 2020                                                                           Wannes Meert
                                                                                        Marian Verhelst

# Acknowledgements

We would like to acknowledge several people and institutions for their valuable contributions to this work.

We thank Nilesh Ahuja, Herman Bruyninckx, Jesse Davis, Guy Van den Broeck, and Patrick Wambacq for their feedback on previous versions of this manuscript and for all the engaging and inspiring discussions throughout the last years. We would also like to acknowledge our collaborators' contributions in the articles discussed throughout this book: Martin Andraud, Komail Badami, Steven Lauwereins, and Nimish Shah.

Finally, we thank Intel for their financial support, as well as the European Research Council for their funding under the Re-SENSE project grant number ERC-2016-STG-71503.

# Contents

1 **Introduction** .................................................................... 1
   1.1   The IoT Paradigm ........................................................ 2
        1.1.1   Moving Intelligence Towards the *Extreme Edge* ............. 3
   1.2   The Machine Learning Pipeline ..................................... 4
        1.2.1   Tasks in Machine Learning .................................... 5
        1.2.2   Performance Evaluation ....................................... 6
        1.2.3   Models .......................................................... 8
        1.2.4   Features ........................................................ 9
   1.3   Resource-Efficient Machine Learning .............................. 10
        1.3.1   Inference ....................................................... 10
        1.3.2   Feature Extraction ............................................. 11
        1.3.3   Learning ....................................................... 12
        1.3.4   Remaining Challenges ........................................ 12
   1.4   Problem Statement ..................................................... 14
   1.5   Sketch of the Proposed Solution ..................................... 15
        1.5.1   The Choice for Probabilistic Models ......................... 15
        1.5.2   System-Wide Hardware-Awareness .......................... 16
   1.6   Structure of This Book ................................................ 18
   References ................................................................... 20

2 **Background** .................................................................... 23
   2.1   Probability Theory..................................................... 24
        2.1.1   Basic Notions and Notation................................... 24
        2.1.2   Probability Distributions ...................................... 25
        2.1.3   Probabilistic Inference ........................................ 26
        2.1.4   Independence Notions and Bayes Rule ...................... 27
   2.2   Bayesian Networks .................................................... 29
        2.2.1   Exact Inference in Bayesian Networks....................... 31
        2.2.2   Parameter Learning ........................................... 32
        2.2.3   Bayesian Network Classifiers ................................ 33

2.3    Probabilistic Circuits ....................................................    34
       2.3.1   Properties of Probabilistic Circuits ...........................    35
       2.3.2   Structural Constraints ..........................................    36
       2.3.3   Classification Tasks with Probabilistic Circuits ..............    37
2.4    Sensory Embedded Pipeline .........................................    38
       2.4.1   Building Blocks .................................................    38
References ....................................................................    39

3   **Hardware-Aware Cost Models** ..............................................    41
    3.1    Hardware-Aware Cost ...............................................    42
    3.2    Sensing Costs ........................................................    43
           3.2.1   Resource Versus Quality Trade-Offs of
                   Mixed-Signal Sensor Front-Ends ...........................    44
    3.3    Feature Extraction Costs ............................................    47
           3.3.1   Digital Feature Precision Scaling ............................    49
           3.3.2   Analog Feature Precision Scaling ...........................    49
    3.4    Inference Costs ......................................................    49
    3.5    Dynamic Tuning Costs for Run-Time Strategies ...................    51
    3.6    Types of Systems Considered in This Book ........................    52
    3.7    Conclusion ...........................................................    53
    References ....................................................................    53

4   **Hardware-Aware Bayesian Networks for Sensor Front-End**
    **Quality Scaling** ...............................................................    55
    4.1    Noise-Scalable Bayesian Network Classifier .......................    56
           4.1.1   Model ............................................................    56
           4.1.2   Inference .......................................................    58
    4.2    Local Pareto-Optimal Feature Quality Tuning .....................    58
    4.3    Use Cases of the ns-BN: Introduction .............................    61
    4.4    First Use Case: Mixed-Signal Quality Scaling ....................    62
           4.4.1   Experiments for Mixed-Signal Quality Scaling ..............    63
    4.5    Second Use Case: Digital Quality Scaling .........................    68
           4.5.1   Experiments for Digital Quality Scaling ....................    69
    4.6    Third Use Case: Analog Quality Scaling ...........................    71
           4.6.1   Experiments for Analog Quality Scaling ....................    72
    4.7    Related Work ........................................................    75
    4.8    Discussion ...........................................................    77
    References ....................................................................    78

5   **Hardware-Aware Probabilistic Circuits** ...................................    81
    5.1    Preliminaries ........................................................    82
           5.1.1   Probabilistic Sentential Decision Diagrams .................    83
    5.2    Hardware-Aware System-Wide Cost .............................    85
    5.3    Pareto-Optimal Trade-off Extraction ..............................    86
           5.3.1   Search Strategy ................................................    87
           5.3.2   Pareto-Optimal Configuration Selection .....................    88

5.4   Experiments: Pareto-Optimal Trade-off .............................. 89
      5.4.1   Embedded Human Activity Recognition...................... 90
      5.4.2   Generality of the Method: Evaluation on
              Benchmark Datasets ......................................... 93
5.5   Learning PSDDs with a Discriminative Bias for
      Classification Tasks .................................................. 95
      5.5.1   Discriminative Bias for PSDD Learning ..................... 96
      5.5.2   Generative Bias and Vtree Learning ......................... 101
5.6   Experiments: Biased PSDD Learning ............................... 101
      5.6.1   Experimental Setup .......................................... 101
      5.6.2   Evaluation of D-LearnPSDD ................................. 102
      5.6.3   Impact of the Vtree on Discriminative Performance ......... 103
      5.6.4   Robustness to Missing Features ............................. 105
5.7   Related Work ......................................................... 106
5.8   Discussion ........................................................... 107
References ................................................................... 108

**6   Run-Time Strategies** ..................................................... 111
6.1   Noise-Scalable Bayesian Networks: Missing Features at Run-Time   112
      6.1.1   Run-Time Pareto-Optimal Selection ......................... 113
      6.1.2   Experiments: Robustness to Missing Features of ns-BN ..... 116
      6.1.3   Digital Overhead Costs ...................................... 118
      6.1.4   Remaining Challenges ....................................... 119
6.2   Run-Time Implementation of Hardware-Aware PSDDs:
      Introduction .......................................................... 119
6.3   Model Complexity Switching Strategy ............................... 120
      6.3.1   Model Selection .............................................. 121
      6.3.2   Model Switching Policy ...................................... 122
      6.3.3   Time Aspects ................................................ 126
      6.3.4   Experiments for Model Switching: Introduction ............. 128
      6.3.5   Model Selection for Experiments ............................ 129
      6.3.6   Hyperparameter Selection ................................... 130
      6.3.7   Performance of the Model Switching Strategy ............... 132
      6.3.8   Robustness to Missing Features ............................. 132
6.4   Related Work ......................................................... 134
6.5   Discussion ........................................................... 136
References ................................................................... 137

**7   Conclusions** ........................................................... 139
7.1   Overview and Contributions ......................................... 140
7.2   Suggestions for Future Work ........................................ 141
7.3   Closing Remark ...................................................... 142
Reference .................................................................... 143

**A   Features Used for Experiments** .............................................. 145
    A.1   Synthetic Dataset for Digital Scaling ns-BN ......................... 145
    A.2   Six-Class HAR Classification with ns-BN ........................... 145

**B   Full List of Hyperparameters for Feature Pruning** ...................... 147

**C   Full List of Hyperparameters for Precision Reduction** .................. 153

**Index** .............................................................................. 159

# Chapter 1
# Introduction

The last decades have witnessed an explosion in the number of electronic devices populating our environment. As foreshadowed by Gordon E. Moore in 1965 [1], this phenomenon owes to the steady decrease in computing costs facilitated by semiconductor technology scaling. Today, electronic devices are ingrained in every aspect of our lives. From the moment we wake up to our smartphone alarms, through daily commutes guided by our cars' navigation systems, to the end of the day, relaxing to a good book loaded on our e-readers.

The computing advancement enabling such a scene of device ubiquity has also inspired the proposal of visionary technological paradigms such as the Internet-of-Things (IoT). Kevin Ashton devised the IoT concept in 1999 in an effort to link the *then* new idea of RFID to the *then* hot topic of the Internet [2]. He argued in favor of a centralized and remote processing architecture exploiting the Internet, where electronic devices would be endowed with the autonomy to gather and share data, thus releasing human beings from the tediousness of performing these tasks manually. Over the following years, this insightful idea would take a life of its own and, up to this day, attempts to realize it would guide many business and research efforts [3, 4].

One of the byproducts of the technological revolutions described above is the tremendous amount of data available today, gathered by the devices connected to the IoT and by the ubiquitous mass of personal computers and smartphones. The so-called *Global Datasphere*, comprising the summation of all the data created, gathered, stored, and replicated, is predicted to grow to 175 zettabytes by 2025 [5]. Data has, in fact, become a boon for algorithmic innovation, especially in the case of machine learning, which feeds off of it [6, 7]. With data increasingly becoming a commodity and the ever demanding constraints of technological applications, the interest in endowing IoT and other battery-powered mobile devices with complex processing capabilities is rapidly rising. Think, for example, of self-driving cars with critical real-time constraints or the legitimate privacy concerns associated with the use of health trackers. Naturally, the main challenge to overcome in this pursuit

L. I. Galindez Olascoaga et al., *Hardware-Aware Probabilistic Machine Learning Models*, https://doi.org/10.1007/978-3-030-74042-9_1

is the heavily constrained energy budget from this battery-based operation and the limited size budget, which made the devices appealing to users in the first place.

This book attempts to address some of these challenges by proposing machine learning models that are aware of the properties of the hardware hosting them. The main goal is to aid battery-powered devices in striking a successful balance between their limited resource budget and their task performance requirements. This chapter sets the stage for the contributions in this book, reviewing the current state of the *Internet-of-Things* paradigm in Sect. 1.1 and providing a brief overview of the ingredients needed to build *machine learning* systems in Sect. 1.2. The motivation behind realizing resource-efficient machine learning systems is then discussed in Sect. 1.3, as well as the existing solutions and their limitations. Section 1.4 formalizes the problems addressed by this book, and Sect. 1.5 then provides a general sketch of the proposed solutions. Finally, Sect. 1.6 specifies which chapters introduce each of these solutions and lists the previously published work this book relies on.

## 1.1   The IoT Paradigm

One of the main contributions of the IoT is providing a computational framework that spans an extensive range of devices and services, with diverse sensing, comput-ing, and communication roles. This is perhaps better reflected by the vast application promise in McKinsey & Company's IoT definition from 2010 [3]: "sensors and actuators embedded in physical objects—from roadways to pacemakers—are linked through wired and wireless networks, often using the same Internet Protocol (IP) that connects the Internet."

Such a massive scheme implies that nearly any device and service can be connected to the IoT. One way of looking at the IoT is as a paradigm that provides useful categorizations of available computing and communication infrastructure. In general, three layers tend to be identified: the *cloud*, the *fog*, and the *edge*, as shown in Fig. 1.1, which also provides examples of the devices found at each level [8].

In the traditional IoT paradigm, sensor-equipped edge devices such as drones, smart watches, smart sensors, and wearable health patches [9] gather data produced

**Fig. 1.1** Structure of the Internet of Things with examples of devices per level. Based on [8]

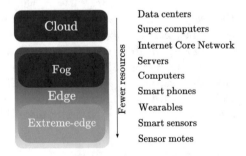

Data centers
Super computers
Internet Core Network
Servers
Computers
Smart phones
Wearables
Smart sensors
Sensor motes

by the environment and the users. Then, they do some light pre-processing on this data, and they finally send it to the cloud [10], where a pool of powerful processing and storage resources is available. The size and, consequentially, the limited computing power of the (usually) battery-powered edge devices motivate this delegation scheme. This deluge of high-quality sensory data has resulted in unprecedented algorithmic innovation, often relying on machine learning models that get increasingly complex in their efforts to advance the state-of-the-art [11, 12]. The infrastructure that has enabled such a centralized cloud-computing approach is today relied upon by many businesses and particular users [5] and continues to be improved upon and showcased [13].

However, complementary approaches to this cloud-centric deployment are currently gaining traction, as they seek to address privacy, latency, and bandwidth constraints through local processing and storage. Within this trend, one can identify two general approaches: fog computing, where cloud-like resources hosted by local servers connect via Ethernet to the IoT edge devices [8] and can perform local processing but also delegate tasks to the cloud; and edge computing, where the devices themselves are capable of performing all processing tasks autonomously but can also decide to deploy some to the cloud.[1]

## 1.1.1 Moving Intelligence Towards the Extreme Edge

The edge layer of the IoT comprises a variety of devices with different processing resources. At the bottom of this layer, or at the *extreme edge*, devices are powered by batteries and therefore have limited computational and storage capabilities. Moreover, they tend to be application-specific. Examples of these smart portable devices are the Google Glass [16], Imec's health patch [17], or Swimtraxx's swim tracker [18]. These devices are application-specific and resource-constrained and often house a System on Chip (SoC) that includes a Central Processing Unit (CPU), memory, and input/output ports that interface with sensors.

Even though several of these devices are equipped to delegate heavier workloads to the cloud, to a local server, or to a mobile device with higher computational capacity, certain applications present requirements that motivate extreme-edge computing, where computation and storage take place fully on-device:

- **Low latency.** Low latency is critical for applications that rely on real-time actions or responses such as autonomous vehicles [19], mobile gaming [15], activity tracking for vulnerable populations [16], remote health monitoring [20], and industrial control systems.
- **Bandwidth intensive applications.** Surveillance and public safety applications often rely on high-bandwidth video streams. Sending them to a remote location

---

[1]Note that other definitions of edge computing contemplate local servers as the main computing element [5, 14], or consider fog computing to be a use case of edge computing [15].

can result in inefficient energy consumption [21]. Furthermore, this can increase latency and can negatively affect their trustworthiness [22].

- **Device location and mobility.** The performance of applications that rely on sensor measurements that change over time (such as velocity and acceleration) can degrade with the latency associated with cloud computing. Such is the case for roadside collision prevention applications [23]. Other applications also benefit from the information generated by user mobility and environmental mobility, especially those that include virtual reality and augmented reality components [15].

State-of-the-art instances of the applications above often rely on machine learning to bring intelligence to the devices found at all different levels of the IoT [24–26]. Machine learning is a sub-field of Artificial Intelligence (AI) that focuses on the extraction of useful patterns from data with the goal of improving the task-specific performance of the targeted application or system as more data or feedback is provided. The main trait of machine learning is that it provides systems with the ability to learn skills and concepts without programming them in advance. The systems can then use this acquired knowledge to perform a task on new data through a process called inference.

Resource efficiency is crucial for implementing machine learning in extreme-edge devices due to their limited computational and storage capabilities. This is particularly important for the inference stage, as it must execute the task of interest at run-time, relying on the resources available within the device. The learning stage, during which the skills and concepts mentioned above are derived, often takes place at an off-line stage, with access to more varied and powerful resources. Nevertheless, learning involves computationally expensive optimization algorithms and tends to rely on large volumes of data. Resource efficiency in this stage is also a relevant problem, whether it takes place on-device or off-device.

The challenges associated with extreme-edge machine learning in terms of energy efficiency are discussed in Sect. 1.3. First, it is necessary to understand the steps involved in implementing a machine learning solution, as well as the different types of models available and the different tasks that can be fulfilled. The following section provides an overview of the steps a machine learning system follows and a general categorization of the different approaches.

## 1.2   The Machine Learning Pipeline

Machine learning addresses *tasks* by mapping input data, described by *features*, to outputs as shown in Fig. 1.2. Moreover, the mapping, or *model*, is itself the output of a learning problem, which relies on a learning algorithm to find the appropriate parameters from *training data*.

The high-level pipeline in Fig. 1.2 highlights three of the main aspects of machine learning: the types of tasks it can address; how to extract information—in the form

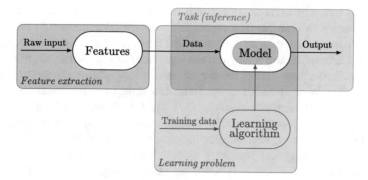

**Fig. 1.2** An overview of the machine learning pipeline. Based on [6]

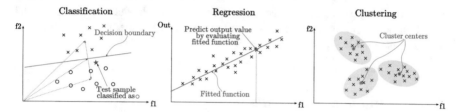

**Fig. 1.3** Examples of classification (based on [6]), regression, and clustering tasks

of features—that is relevant to the task of choice; and how these two aspects affect the model selection process.

## 1.2.1   Tasks in Machine Learning

One of the most common tasks in machine learning is classification, where the goal is identifying which category a new observation belongs to. Consider, for example, an activity recognition application where the user carries a wearable equipped with an accelerometer and a gyroscope. The goal of a *binary classification* task could be to identify whether the person is running or walking or whether the person is walking, running, climbing stairs, or standing, for a *multi-class classification* task. In this type of task, the model in Fig. 1.2 maps the raw input accelerometer and gyroscope signals to features and then to the available output classes. The leftmost part of Fig. 1.3 illustrates a two-dimensional feature space and a binary classification task, where a *decision boundary* partitions the examples according to their labels.

Classification assumes a pre-defined set of categories or classes. In some contexts, however, predicting a real number may be more desirable. Suppose the wearable in the example above is used in the context of elderly care. A healthcare provider could, for example, evaluate whether the patient is suffering from cognitive decline by observing the level of difficulty of the activities performed on a sliding scale. This task is called regression, and it maps input data to a real value. Figure 1.3 (center) shows an example of a regression task, where a function is fitted to the available data and can then be used to predict the output value of a new example.

Both classification and regression are *supervised* tasks, where a set of examples labeled with the true classes or output values is available. However, labeling examples is a labor-intensive process and can sometimes be impossible, like in the case of astronomical observations [27]. Machine learning tasks that lack prior information about class membership or output values are called *unsupervised*. The prime example of unsupervised classification is called *clustering*, where data is grouped by assessing the similarity among the available instances, like in the rightmost portion of Fig. 1.3, where samples are grouped based on the distance to each other. In the case of activity recognition, a clustering algorithm would be capable of identifying that inputs with large values of acceleration in the vertical direction belong to one type of activity. In contrast, large values in the horizontal direction belong to another one. This without a priori knowledge on the types of activities the user is capable of performing.

## 1.2.2   Performance Evaluation

The performance of machine learning algorithms is evaluated in terms of how well they can execute the task they were *trained* for on a new set of data. Therefore, the available data is split (often randomly) into a training set (e.g. 80%) and a testing set (the remaining 20%). The training process aims to find useful patterns in the training data, producing a model capable of solving the desired task autonomously. Figure 1.3 and the top of Fig. 1.4 illustrate such a pattern or *decision boundary*, meant to be deployed for a binary classification task.

Classification performance is often measured by counting the number of correctly classified samples and dividing it by the total number of available samples, getting a metric called *accuracy*. If the machine learning model fails in generalizing the learned patterns or concepts to a new situation, as shown in Fig. 1.4, accuracy may be low. The model may be too simple to describe the training data's variance, leading to under-fitting and low accuracy, both for the train and test sets. On the other hand, the model may be too complex and over-fit to the training data, memorizing the patterns present and failing to perform appropriately on unseen new data. In this situation, training accuracy will be significantly higher than testing accuracy.

It is also useful in most classification scenarios to break down the proportion of correctly and incorrectly classified instances according to their class membership and class prediction. Confusion matrices, like the one illustrated at the bottom-left of Fig. 1.4 for a 6-class activity recognition scenario, can provide such an analysis. The rows in this example correspond to the true labels, while the columns correspond to the predictions. Correctly classified samples are identified as true positives (TPs),

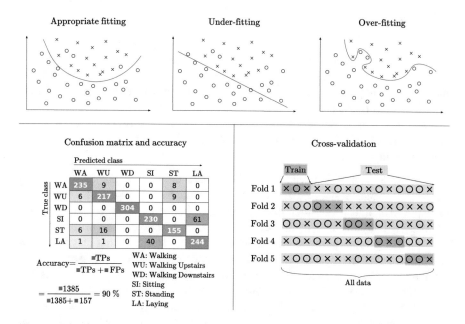

**Fig. 1.4** Top: Examples of appropriate fitting, over-fitting, and under-fitting in a binary classification task. Bottom-left: Example of a confusion matrix for 6-class activity recognition. Bottom-right: Process of cross-validation over five-folds

while incorrectly ones as false positives (FPs).[2] The confusion matrix shows that the classifier has some difficulties in identifying *sitting* from *laying*, while it has an easier time correctly identifying *walking downstairs*.

Finally, note that one can occasionally be lucky or unlucky in the random train–test split. If the sampled test set happens to be noisy, the model might appear to underperform. If the test set samples are very similar to those in the train set, the model will attain high accuracy. Therefore, the splitting is often repeated in a process called *cross-validation*. The data is randomly divided into $k$ equally sized parts, $k - 1$ are used for training, and the remaining for testing. This process is performed $k$ times in a sliding fashion, as shown at the bottom-right of Fig. 1.4, and the final accuracy is obtained by averaging the test set results over the $k$ folds.

---

[2]The terminology is simplified to describe a general case of multi-class classification. In binary classification, one of the class labels can be deemed *positive* and the other *negative*. In this situation, the terms true negative (TN) and false positive (FP) are also relevant.

## 1.2.3   Models

Models are the main output of machine learning, as they learn to perform the desired task from data, like in the examples from the previous section. The range of tasks machine learning can address is considerable; consequently, the range of available models is equally sizable.

Peter Flach proposes a (non-mutually exclusive) categorization of machine learning models into three general groups: geometric models, probabilistic models, and logical models [6].

**Geometric Models** This type of models is built directly in feature space using geometric concepts such as lines, planes, and distances. Figure 1.3 shows an example of a linear classifier that constructs a boundary between the × and ○ samples by half-way intersecting the line between the two centers of mass. Other examples of geometric models are nearest-neighbor classifiers that classify a new instance by evaluating its Euclidean distance and assigning it to the same class as the closest training instance. Support vector machines learn a boundary that has maximum distance to the closest training example of each type of class. Finally, several types of neural networks, such as the multi-layer perceptron, are also based on geometric concepts, as they attempt to find a non-linear hyper-plane that separates instances from the different classes in feature space.

**Probabilistic Models** These models assume that there is an underlying random process generating the variables (or features) and that a probability distribution can therefore describe a relation between them. Tasks can be performed with these models using probabilistic axioms and rules. For the running example on activity recognition, one can infer the probability of the possible activities given sensory data by learning a distribution over the available features and using Bayes rule[3] to calculate the posterior probability as shown in Fig. 1.5. That is, for example, by asking *what is the probability that I am jumping or walking given that my accelerometer reading in the vertical direction has a high variance and the one in the horizontal direction has a fairly low mean value?* Moreover, the representation of probabilistic models often implies that they are generative, meaning that one can sample values from the variables included in the model.

**Logical Models** Models of this type rely on a sequence of logical rules to perform the task of interest. This sequence is commonly organized in a tree or graph structure, like in the case of decision trees, as shown in Fig. 1.5. Here, the decision tree partitions the feature space hierarchically, and the leaves are labeled with the different classes. Classification of a new instance in this example can be performed

---

[3]Recall that Bayes theorem describes the probability of an event given prior knowledge and observations and is given by $\Pr(Y|X) = \frac{\Pr(X|Y)\Pr(Y)}{\Pr(X)}$, where $\Pr(X|Y)$ is the likelihood function and $\Pr(Y)$ is the prior probability.

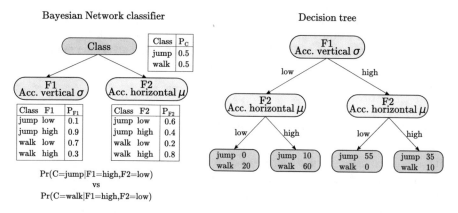

**Fig. 1.5** Examples of Bayesian network classifier and a decision tree

by following the path of the tree in accordance to the encoded rules until the leaf level, and predicting the majority class of that particular leaf.

As mentioned before, the categorization above is not mutually exclusive. For instance, Probabilistic Circuits, the representation that the contributions in Chap. 5 rely on, combine notions from probabilistic and logical models. Similarly, one can find analogies in the loss functions used for learning the different types of models.

## 1.2.4 Features

Features are measurable properties of the input data and are essential to the success of machine learning tasks, since models are defined in terms of them. In most practical scenarios, features need to be constructed by the developer or the machine learning application. For example, activity recognition applications often rely on statistical quantities extracted from the filtered and windowed raw sensory data, such as the mean or standard deviation of a windowed acceleration signal. Features are used in lieu of the raw signal because the latter can be sparser and noisier.

Practical scenarios often demand an iterative process [7] of feature extraction → model training → model evaluation → feature processing, modification or additional extraction → model training → ... Two common approaches to modify feature sets in this iterative process are discretization and feature selection. Discretization reduces the domain of a feature by grouping the available values into pre-defined bins. This is often used for sparse real-valued features. For example, a feature expressing the height of a small group of people may benefit from a scale that increases every five centimeters, instead of a scale that increases every millimeter. Another situation where sparsity may have a negative effect on the success of the machine learning algorithm are high-dimensional feature spaces.[4] In this situation, feature selection can aid in reducing the size of the feature space and selecting only those that are relevant for the task. Moreover, this can prevent the model from becoming unduly complex and over-fitting to the training data.

## 1.3  Resource-Efficient Machine Learning

It is clear from Sect. 1.2 that each stage of the machine learning pipeline is essential for the success of the task. Naturally, each stage poses a unique set of challenges that are exacerbated when deployed in a resource-constrained environment, as discussed in Sects. 1.3.1, 1.3.2, and 1.3.3 for the inference, the feature extraction, and the learning stages of the pipeline, respectively. These sections also list the existing solutions and related work, but Sect. 1.3.4 illustrates some of the remaining challenges not yet addressed by them.

### 1.3.1  Inference

Battery-powered edge devices have strict energy budget constraints, often functioning in power consumption regions of <100 mW [28]. Size and cost constraints also limit the amount of available memory and compute power. These capabilities are in stark contrast with state-of-the-art machine learning implementations, whose inference stage can require tens of billions of computations per second and gigabytes of storage space [11].

Such a demanding workload is not attainable at the extreme edge, where devices are equipped with embedded CPUs that can only perform at ranges lower than ten giga operations per second, and only include very small memories or can be equipped with external ones of up to a few megabytes [29]. This challenge, however, has not stopped the unrelenting pursuit of machine learning at the extreme edge, for the reasons postulated in Sect. 1.1.1. In turn, it has driven the emerging fields of resource-efficient machine learning and specialized edge-based AI hardware [30].

Among the large variety of machine learning models (see Sect. 1.2.3), Deep Neural Networks (DNNs)[5] have arguably been most successful at leveraging the exponential compute progress brought about by the democratization of electronic devices [31] and have undeniably been able to deliver impressive results in a variety of applications such as speech recognition, vision, natural language processing,

---

[4]This is often referred to as the *curse of dimensionality* [6].

[5]Deep neural networks are a range of machine learning models that map input data to outputs through connected and weighted layers of non-linear functions and are capable of learning highly complex hyper-planes in feature space.

and robotics [28, 32]. Therefore, today they are considered the de facto machine learning model by many industrial and academic actors, including those interested in extreme-edge computing and custom hardware [33–35]. Consequently, most extreme-edge realizations of machine learning are currently located in the realm of DNNs, both from the algorithmic and from the hardware point of view. One can identify the following trends in the pursuit of efficient inference on DNNs at the extreme edge [28]:

- Model topology modification and pruning: A common strategy consists of constructing the most compact possible model by "pruning" redundant connections and weights and exploiting structural and weight sparsity [36].
- Parameter quantization: Reducing the precision of parameters often comes with significant resource consumption savings and small accuracy losses, acceptable for most applications [37].
- Specialized "neural" hardware: DNNs lend themselves to parallelization due to their layered and uniform structures. The use of Graphics Processing Units (GPUs) has exploited these parallelization opportunities and other properties. However, more recent trends have opted for custom processors optimized for the workloads of the DNN, such as the Neural Processing Unit (NPU) [38].
- Emerging memory-based strategies: Memory transactions are perhaps the costliest operation in embedded hardware [39]. Emerging strategies target the reduction of such transactions by exploiting the properties of memory cells (most commonly SRAM) to perform local computing before engaging in unnecessary transactions [40].

The non-exhaustive list above serves as a sampler of the wide variety of strategies machine learning developers have devised to improve the resource efficiency of performing inference in complex models.

## 1.3.2 Feature Extraction

The feature extraction stage is not immune to the constraints of extreme-edge devices. Together with inference, feature extraction must be deployed at runtime in most portable applications equipped with sensors, as the incoming signal must be mapped to a representation useful for the model. The constraints in this process have also been addressed by algorithmic and hardware techniques. Feature selection strategies are sometimes augmented with resource-aware cost functions that represent, for example, the energy consumption associated with extracting the different types of features [41, 42]. On the hardware design front, one trend has been to reduce computation overhead by attempting to extract information as close as possible to the sensor itself using analog feature extraction [43] or through *analog-to-information converters* [44, 45].

### 1.3.3 Learning

Most machine learning implementations rely on a resource inefficient model training process. This process tends to involve computationally expensive optimization algorithms and often depends on large volumes of data for convergence [46, 47]. Attempts to address resource efficiency during this stage rely on techniques similar to the ones referenced above, such as exploiting parallelism and reducing precision [48] and targeting distributed learning approaches at the edge, such as in the case of federated learning [49].

The main bottlenecks towards always-on machine learning at the extreme edge are still found at the stages that must take place at run-time, namely feature extraction and inference. Therefore, this book focuses on these stages' resource efficiency challenges and assumes that some of the burdens can be delegated or off-loaded to the off-line training stage.

It follows from this observation that the learning stage can also play an important role in ensuring resource efficiency at inference time. A relevant body of work targeting this aspect is cost-sensitive learning [50], where different types of costs are taken into consideration for the process of decision making. A common approach within this area is to trade-off the cost of acquiring a new feature in the process of decision making, and the resulting prediction performance, either optimally at an off-line stage or at run-time relying on heuristics [51–54]. Another approach is to rely on progressively costlier models and accept a prediction based on a desired confidence level [55, 56].

It is clear that the distinct steps necessary to fulfill a machine learning task are implemented in different sub-systems of the extreme-edge device. Consequently, they suppose different requirements and challenges that have been addressed by a vast range of resource-aware solutions, as evidenced throughout Sect. 1.3. However, most of these solutions target particular stages of the pipeline and miss resource-saving opportunities brought by considering the system as a whole. Furthermore, extreme-edge applications' inherently dynamic conditions pose a number of additional challenges that can be approached with the same integral approach. These remaining challenges are illustrated throughout the following section.

### 1.3.4 Remaining Challenges

The mobile activity recognition scenario in Fig. 1.6 aids in illustrating some of the remaining challenges in the pursuit of intelligence at the extreme edge. The device in this scenario includes three sensors, and two features can be extracted from each. The goal of this application is to classify the user's activity with an unspecified machine learning algorithm given in terms of four features.

Consider the available feature subsets $F_1$, $F_2$, $F_3$, and $F_4$, defined at the top of Fig. 1.6. Suppose that feature subsets $F_1$ and $F_2$ lead to the same classification

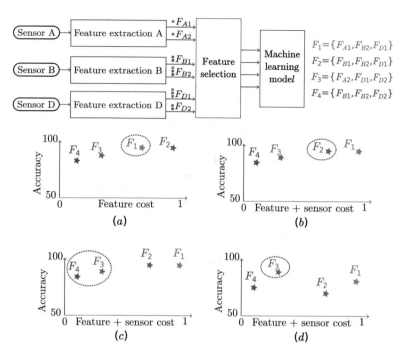

**Fig. 1.6** Sensory embedded classification example illustrating some of the remaining challenges towards the pursuit of resource-efficient always-on functionality

accuracy, as shown in Fig. 1.6a, but that the total feature cost of $\mathbf{F}_1$ is lower than that of $\mathbf{F}_2$. Note that feature cost can, for example, be estimated by counting the required number of arithmetic and memory operations. The problem of resource-aware feature selection consists of looking for the feature subset that maximizes performance and minimizes resource consumption. The solution to this problem in the current scenario is clearly $\mathbf{F}_1$, as highlighted in Fig. 1.6a.

However, this approach disregards the costs incurred by the rest of the system, as shown in Fig. 1.6b. When considering sensor cost in addition to feature cost, subset $\mathbf{F}_2$ turns out to be the better choice, as none of the features in this subset are extracted from sensor $A$ and can be therefore turned off. This example shows that trade-off opportunities can be missed when failing to describe realistic hardware-aware system-wide costs and the scalability available in the system (in this example, the fact that sensor $A$ can be turned off).

The dynamic behavior of run-time scenarios poses additional challenges. Battery life limitations call for on-demand low-cost operation modes. This is the case in Fig. 1.6c, where subsets $\mathbf{F}_3$ or $\mathbf{F}_4$ are preferred when the battery is running low, since they operate in low-cost regions, and exchange these gains with only small accuracy losses that the application and the user may be willing to tolerate. Furthermore, certain dynamic changes in the environment and the device demand real-time responses. Consider Fig. 1.6d where sensor $B$ suddenly malfunctions, but the other two sensors remain functional. In this situation, the optimal cost versus accuracy trade-off is clearly achieved by feature subset $\mathbf{F}_3$, since the system experiences significant accuracy degradation under feature subset $\mathbf{F}_2$, which relies on the failing sensor. Had this feature subset been selected off-line and fixed for run-time implementation, the system would have lost all robustness unless it was enabled to switch to a different subset dynamically as required.

To summarize, the following open challenges are identified in the pursuit of machine learning at the extreme edge:

1. Holistic resource-aware machine learning strategies are required to address the trade-offs induced by the different components and stages of the system.
2. It is necessary to define system costs under a unified—and relevant—resource, such as energy consumption.
3. There is a need to address the dynamic requirements of run-time applications, such as robustness to missing information and changing battery availability.

These problem statements of this book are formulated accordingly in Sect. 1.4.

## 1.4 Problem Statement

The three main problems this book aims to address are detailed below and are based on the open challenges discussed in the previous section.

**Bringing Hardware-Awareness to the Machine Learning Pipeline** Section 1.3.4 illustrates a scenario where not taking into consideration the cost incurred by the full system leads to missed cost-saving opportunities. Therefore, the first problem this book focuses on is how to define hardware-aware cost metrics that can represent several levels of abstraction of a device under a common framework and in terms of a measurable and relevant resource, such as energy consumption.

**Exploiting Hardware-Awareness Towards Optimal Cost Versus Performance Trade-Offs** The fundamental trade-off between cost and accuracy[6] is the driver in all the contributions of this book. Figure 1.6 demonstrates that the optimal accuracy versus cost trade-off is determined by the particular circumstances and objectives of the application. Therefore, the goal of the techniques discussed in this book is to extract the Pareto-optimal front within this trade-off space, such that, for any given cost constraint, accuracy is maximized. Or conversely, such that, for any given accuracy requirement, cost consumption is minimized. Furthermore, this book explores how hardware-awareness can tap into scalability opportunities not targeted by traditional resource-aware strategies, such as reduced sensor quality or scalable model complexity.

---

[6]The task of focus on this book is classification since it is the most commonly deployed by extreme-edge devices.

**Dynamic Selection of Pareto-Optimal Operating Points** The last problem considers the dynamic nature of run-time scenarios. As illustrated in Sect. 1.3.4, setting a static operating point at an off-line stage can result in severe performance degradation at run-time, when the device's functionality can falter unexpectedly. Similarly, the very nature of battery-powered devices implies that the accuracy versus cost needs may change over time. This calls for strategies that are capable of dynamically selecting appropriate cost versus accuracy operating points that can address the instantaneous needs of the user and the application.

Based on the goals describe above, Sect. 1.5 specifies a *wish-list* of the properties guiding the selection of the machine learning model considered in this book. It then provides an overview of the steps followed to make these models hardware-aware.

## 1.5   Sketch of the Proposed Solution

This section offers a sketch of the solutions proposed to address the problems above towards always-on hardware-aware machine learning in extreme-edge devices. First, Sect. 1.5.1 motivates the choice for probabilistic models to realize the hardware-awareness proposed in this book and then Sect. 1.5.2 provides an overview of the steps followed by these strategies.

### 1.5.1   The Choice for Probabilistic Models

Section 1.2 argues that the selection of an appropriate machine learning model must be tied to the task of interest and to the nature of the data procured by the application. Realizing the hardware-awareness vision above adds one more layer to the list of requirements. The reader will witness throughout this book that probabilistic models are equipped to meet most of these requirements, owing to their versatile properties, briefly discussed below and summarized in Table 1.1. Note that the proofs of concept in this book focus on Bayesian networks and Probabilistic Circuits as they constitute two representative examples of probabilistic models:

- The dynamic situations described above, where features may be unavailable, can be handled by probabilistic models, since they can be generative and allow

**Table 1.1** Enabling properties of probabilistic models towards hardware-awareness.

| Problem/goal | Properties of probabilistic models |
| --- | --- |
| Robustness to missing features | They are generative |
| User-friendly and ease of debugging | Can be interpretable |
| Encode hardware properties in the model | Allow to represent expert knowledge |
| Induce cost vs. accuracy trade-offs | State-of-the-art models do this while training |

to perform inference even when certain variables are not available. Moreover, the hardware-awareness strategies that this book discusses also exploit this to represent scenarios where features or sensors are willingly turned off to save cost.

- Some of the hardware properties of the extreme-edge devices targeted by this book can be represented with probability distributions, such as in the case of sensor noise. Probabilistic models enable the incorporation of expert knowledge, and therefore, the hardware properties of interest can be directly encoded in the model.
- The hardware-aware strategies discussed in this book attempt to exploit the scalable properties of the device, from the quality of the sensors streams, to the type and number of extracted features, to how complex the model is. In a practical scenario, the user may want to understand why the settings of their system change in real time, and the developer may want to debug actions related to this hardware-aware behavior. Unlike other types of models (specially the geometric ones), probabilistic models can be interpretable and can thus provide the user and the developer with a good understanding of their behavior.
- Finally, the cost versus accuracy trade-off that this book aims to optimize is explicitly enforced by certain state-of-the-art probabilistic models, which take such a trade-off into consideration already from the training stage.

### 1.5.2  System-Wide Hardware-Awareness

Endowing probabilistic models with hardware-awareness takes place over the following stages, also illustrated by Fig. 1.7:

1. **Identifying scalable properties of the target hardware.** The top of Fig. 1.7 shows an example of three scalable properties in the activity recognition example. The device in this example allows to scale the quality of the sensory signal and the extracted features (highlighted in purple), as well as which features can be extracted and how many are available at any given time (highlighted in green). Finally, the complexity of the probabilistic model in this example can be tuned to be higher (and more expressive and costly) or lower (and simpler but cheaper) or to encode its parameters and perform arithmetic operations with more or less bits, as highlighted in blue.
2. **Mapping these scalable properties to the hardware-cost versus accuracy trade-off space**. Each combination of the available scalable properties maps to a different operating point in the cost versus accuracy trade-off space. For example, in Fig. 1.7a, using all sensors and extracting all available features incur the highest possible hardware-cost, defined accordingly, in terms of signal quality and number of features. The accuracy attained by this *highest quality configuration* constitutes the baseline in the trade-off space. Turning off sensors or extracting fewer features will result in less costly operating points at the

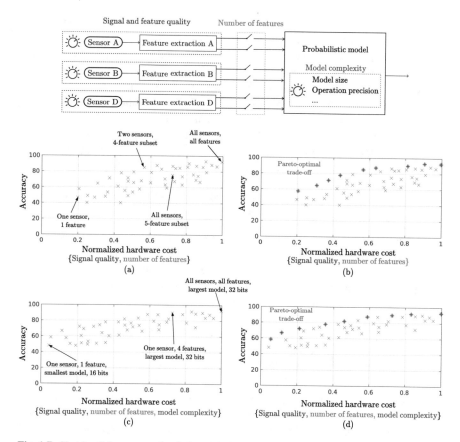

**Fig. 1.7** Sketch of the proposed solution. (**a**) Cost versus accuracy trade-off from feature and sensor scaling. (**b**) Local Pareto-optimal front from feature and sensor scaling. (**c**) Trade-off from model complexity scaling in addition to feature and sensor scaling. (**d**) Local Pareto-optimal from the trade-off in (**c**)

expense of some accuracy losses. Similarly, Fig. 1.7c considers a scenario where not only the sensor and feature extraction portion of the system can be scaled, but also the complexity of the probabilistic model, leading to additional cost savings.

3. **Extracting the Pareto-optimal front**. Figure 1.7b shows the local[7] Pareto-optimal front of the trade-off enabled by the search illustrated in (a). Note, for example, that the *two sensors, 4-feature subset* operating point is part of the Pareto front, whereas the *all sensors, 5-feature subset* is not. The former point achieves the highest reachable accuracy for a normalized cost of 0.58, while the latter achieves a lower accuracy at an even higher cost. A similar situation can

---

[7]The Pareto-optimal fronts throughout this book are extracted from the trade-off space made available by greedy searches. Therefore, they are not globally Pareto-optimal, as exhaustively exploring the full trade-off space is computationally infeasible in most practical cases.

be observed in (c) and (d) where the *one sensor, 4 features, largest model, 32 bits* point is part of the Pareto-optimal. These Pareto fronts can be deployed to determine a configuration of the device that, for a given desired cost, will maximize accuracy or vice versa.

4. **Deploying the Pareto-optimal front at run-time.** When the application's requirements change over time, like in the case of low battery or sensor failure illustrated in Sect. 1.4, the system can benefit from dynamically traversing the Pareto-optimal front, as opposed to selecting a unique point. For example, in Fig. 1.7d, the device might start performing the desired task while under the operating point that attains 95% accuracy at a cost of 81% with respect to the baseline. However, the moment the battery is critically low, it is worth changing to a point that, for example, consumes 40% of the baseline cost at the expense of 15% accuracy loss, than risking the device turning off.

The overarching goal of the techniques discussed in this book is to create opportunities for the efficient implementation of probabilistic models in resource-constrained, extreme-edge devices. This goal encompasses the aspects described in the problem statement and is motivated by the favorable properties of probabilistic models, as described above.

## 1.6   Structure of This Book

The structure of this book follows the sequence of steps described in the previous section. As it progresses, it targets different aspects of hardware scalability, for which it exploits the properties of two types of probabilistic models: first, Bayesian network classifiers and Probabilistic Circuits second. Below is a brief overview of the contents of each chapter, along with a list of related publications.

**Chapter 2** reviews the foundations of probability theory and probabilistic models and provides an overview of the properties of the *embedded sensing pipeline* considered in this book.

**Chapter 3** introduces the notion of hardware-aware cost that guides the optimization frameworks to follow. This notion specifies how a certain *device configuration*, given in terms of the feature and sensor quality and model complexity (like in Sect. 1.5.2), can be mapped to a probabilistic model and how it can be calculated in terms of a measurable hardware resource such as energy consumption. This chapter is partly based on the following publications:

Galindez, L., Badami, K., Vlasselaer, J., Meert, W., and Verhelst, M. (2018). Dynamic Sensor-Frontend Tuning for Resource Efficient Embedded Classification. *IEEE Journal on Emerging and Selected Topics in Circuits and Systems*, 8(4), 858–872.

Galindez Olascoaga, L. I., Meert, W., Shah, N., Verhelst, M. and Van den Broeck, G. (2019). Towards Hardware-Aware Tractable Learning of Probabilistic Models. *In Advances in Neural Information Processing Systems (NeurIPS)* (pp. 13726–13736).

**Chapter 4** presents the *noise-scalable Bayesian network classifier*, a probabilistic model that represents the different levels of signal quality procured by available sensors and their front-ends. It showcases the functionality of this model on three use cases, each targeting a different source of hardware noise, and demonstrates how the model enables the extraction of a (local) Pareto-optimal hardware-cost versus accuracy trade-off, as illustrated in Fig. 1.7a and b. This chapter is based on the following publications:

Galindez Olascoaga, L. I., Meert, W., Bruyninckx, H., and Verhelst, M. (2016). Extending naive Bayes with Precision-Tunable Feature Variables for Resource-Efficient Sensor Fusion. *In 2nd AI-IoT Workshop collocated with ECAI* (Vol. 1724, pp. 23–30). CEUR-WS.

Galindez Olascoaga, L. I., Badami, K., Pamula, V. R., Lauwereins, S., Meert, W., and Verhelst, M. (2016). Exploiting System Configurability towards Dynamic Accuracy-Power Trade-offs in Sensor Front-Ends. *In 50th Asilomar Conference on Signals, Systems and Computers* (pp. 1027–1031). IEEE.

Galindez, L., Badami, K., Vlasselaer, J., Meert, W., and Verhelst, M. (2018). Dynamic Sensor-Frontend Tuning for Resource Efficient Embedded Classification. *IEEE Journal on Emerging and Selected Topics in Circuits and Systems*, 8(4), 858–872.

**Chapter 5** introduces a hardware-aware optimization technique for Probabilistic Circuits, a state-of-the-art deep probabilistic model that enables efficient inference and that can explicitly trade-off expressiveness and complexity at learning time. This hardware-aware approach is equipped to induce the cost versus accuracy trade-off brought about by the hardware scaling at different levels of abstraction of the system, as illustrated in Fig. 1.7c and d. The chapter then introduces a technique for learning Probabilistic Circuits that are robust to missing features (since they are generative models) while improving accuracy on classification tasks (due to the encoding of a *discriminative bias*). These contributions are based on the following publications:

Galindez Olascoaga, L. I., Meert, W., Shah, N., Verhelst, M. and Van den Broeck, G. (2019). Towards Hardware-Aware Tractable Learning of Probabilistic Models. *In Advances in Neural Information Processing Systems (NeurIPS)* (pp. 13726–13736).

Galindez Olascoaga, L. I., Meert, W., Shah, N., Van den Broeck, G., and Verhelst, M. (2020). Discriminative Bias for Learning Probabilistic Sentential Decision Diagrams. *In International Symposium on Intelligent Data Analysis (IDA)* (pp. 184–196). Springer, Cham.

**Chapter 6** builds on the hardware-aware models proposed in Chaps. 4 and 5 and focuses on two challenges posed by run-time scenarios: robustness to missing features and the need for low-cost operation. The first part demonstrates how the Pareto-optimal cost versus accuracy trade-off derived with the techniques in Chap. 4 can be used in a run-time scenario and remain robust to missing features due to failing sensors; and the second one proposes a dynamic model complexity switching strategy that builds on the trade-off derived with the techniques in Chap. 5, demonstrating beyond-Pareto performance and also remaining robust to missing features from failing sensors. The first part of this chapter is based on the following publication:

Galindez, L., Badami, K., Vlasselaer, J., Meert, W., and Verhelst, M. (2018). Dynamic Sensor-Frontend Tuning for Resource Efficient Embedded Classification. *IEEE Journal on Emerging and Selected Topics in Circuits and Systems*, 8(4), 858–872.

While an initial version of the second has been included in the following publication:

Galindez Olascoaga, L. I., Meert, W., Shah, N., and Verhelst, M. (2020b). Dynamic Complexity Tuning for Hardware-Aware Probabilistic Circuits. *In IoT, Edge, and Mobile for Embedded Machine Learning (ITEM) Workshop, collocated with ECML-PKDD 2020.*

**Chapter 7** summarizes the contents of this book, discusses its contributions and their implications, and proposes future research directions.

# References

1. G. Moore, The future of integrated electronics, *Fairchild Semiconductor Internal Publication*, vol. 2 (1964)
2. K. Ashton, That 'Internet of Things' thing (1999)
3. M. Chui, M. Löffler, R. Roberts, The Internet of Things, March 2010 [Online; posted 27-August-2012]. [Online]. Available: https://www.mckinsey.com/industries/technology-media-and-telecommunications/our-insights/the-internet-of-things
4. K. Routh, T. Pal, A survey on technological, business and societal aspects of Internet of Things by q3, 2017, in *2018 3rd International Conference on Internet of Things: Smart Innovation and Usages (IoT-SIU)* (IEEE, 2018), pp. 1–4
5. D. Rydning, J. Reinsel, J. Gantz, The digitization of the world from edge to core. *Framingham: International Data Corporation* (November 2018). [Online]. Available: https://www.seagate.com/files/www-content/our-story/trends/files/idc-seagate-dataage-whitepaper.pdf
6. P. Flach, *Machine Learning: The Art and Science of Algorithms That Make Sense of Data* (Cambridge University Press, 2012)
7. C.M. Bishop, *Pattern Recognition and Machine Learning* (Springer, 2006)
8. J. Portilla, G. Mujica, J.-S. Lee, T. Riesgo, The extreme edge at the bottom of the Internet of Things: A review. IEEE Sensors J. **19**(9), 3179–3190 (2019)
9. A. Al-Fuqaha, M. Guizani, M. Mohammadi, M. Aledhari, M. Ayyash, Internet of Things: A survey on enabling technologies, protocols, and applications. IEEE Commun. Surv. Tutorials **17**(4), 2347–2376 (2015)
10. P. Mell, T. Grance, et al., The NIST definition of cloud computing (2011)
11. A. Canziani, A. Paszke, E. Culurciello, An analysis of deep neural network models for practical applications. Preprint (2016). arXiv:1605.07678
12. C. Li, OpenAI's GPT-3 language model: A technical overview, June 2020. [Online]. Available: https://lambdalabs.com/blog/demystifying-gpt-3/
13. K. Naveen, Google breaks AI performance records in MLPerf with world's fastest training supercomputer. Available at https://cloud.google.com/blog/products/ai-machine-learning/google-breaks-ai-performance-records-in-mlperf-with-worlds-fastest-training-supercomputer(2020/07/29)
14. W. Yu, F. Liang, X. He, W.G. Hatcher, C. Lu, J. Lin, X. Yang, A survey on the edge computing for the Internet of Things. IEEE Access **6**, 6900–6919 (2018)
15. G. Premsankar, M. Di Francesco, T. Taleb, Edge computing for the Internet of Things: A case study. IEEE Internet Things J. **5**(2), 1275–1284 (2018)
16. K. Ha, Z. Chen, W. Hu, W. Richter, P. Pillai, M. Satyanarayanan, Towards wearable cognitive assistance, in *Proceedings of the 12th Annual International Conference on Mobile Systems, Applications, and Services* (2014), pp. 68–81

17. Imec, Disposable health patch. Available at https://www.imec-int.com/en/circuitry-sensor-hubs/disposable-health-patch(2020/08/04)
18. Swimtraxx, The smartest swim specific system ever. Available at https://www.swimtraxx.com/pages/device(2020/08/04)
19. S. Liu, L. Liu, J. Tang, B. Yu, Y. Wang, W. Shi, Edge computing for autonomous driving: Opportunities and challenges. Proc. IEEE **107**(8), 1697–1716 (2019)
20. A.A. Abdellatif, A. Mohamed, C.F. Chiasserini, M. Tlili, A. Erbad, Edge computing for smart health: Context-aware approaches, opportunities, and challenges. IEEE Network **33**(3), 196–203 (2019)
21. A. Kumar, S. Goyal, M. Varma, Resource-efficient machine learning in 2 kb RAM for the Internet of Things, in *International Conference on Machine Learning* (2017), pp. 1935–1944
22. B. Kantarci, H.T. Mouftah, Trustworthy sensing for public safety in cloud-centric Internet of Things. IEEE Internet Things J. **1**(4), 360–368 (2014)
23. D. Anadu, C. Mushagalusa, N. Alsbou, A.S.A. Abuabed, Internet of Things: Vehicle collision detection and avoidance in a VANET environment, in *2018 IEEE International Instrumentation and Measurement Technology Conference (I2MTC)* (2018), pp. 1–6
24. F. Samie, L. Bauer, J. Henkel, From cloud down to things: An overview of machine learning in Internet of Things. IEEE Internet Things J. **6**(3), 4921–4934 (2019)
25. I.U. Din, M. Guizani, J.J. Rodrigues, S. Hassan, V.V. Korotaev, Machine learning in the Internet of Things: Designed techniques for smart cities. Future Gener. Comput. Syst. **100**, 826–843 (2019)
26. H. Li, K. Ota, M. Dong, Learning IoT in edge: Deep learning for the Internet of Things with edge computing. IEEE Network **32**(1), 96–101 (2018)
27. Y. Zhang, Y. Zhao, Automated clustering algorithms for classification of astronomical objects. Astron. Astrophys. **422**(3), 1113–1121 (2004)
28. M. Verhelst, B. Murmann, Machine learning at the edge, in *NANO-CHIPS 2030* (Springer, 2020), pp. 293–322
29. ARM, ARM classic processors (2014). Available at https://developer.arm.com/ip-products/processors/classic-processors
30. S. Hooker, The hardware lottery. Preprint (2020). arXiv:2009.06489
31. R. Sutton, The bitter lesson. Incomplete Ideas (blog) **13**, 12 (2019)
32. V. Sze, Y.-H. Chen, T.-J. Yang, J.S. Emer, Efficient processing of deep neural networks: A tutorial and survey. Proc. IEEE **105**(12), 2295–2329 (2017)
33. J. Dean, 1.1 the deep learning revolution and its implications for computer architecture and chip design, in *2020 IEEE International Solid-State Circuits Conference - (ISSCC)* (2020), pp. 8–14
34. Y. LeCun, 1.1 deep learning hardware: Past, present, and future, in *2019 IEEE International Solid-State Circuits Conference - (ISSCC)* (2019), pp. 12–19
35. P. Warden, D. Situnayake, *TinyML: Machine Learning with TensorFlow Lite on Arduino and Ultra-Low-Power Microcontrollers* (O'Reilly Media, 2019). [Online]. Available: https://books.google.be/books?id=sB3mxQEACAAJ
36. S. Han, H. Mao, W. Dally, Deep compression: Compressing deep neural network with pruning, trained quantization and Huffman coding. *CoRR* (2016)
37. Y. Guo, A survey on methods and theories of quantized neural networks. Preprint (2018). arXiv:1808.04752
38. D.A. Palmer, M. Florea, Neural processing unit, February 2014, US Patent 8,655,815
39. M. Horowitz, 1.1 computing's energy problem (and what we can do about it), in *2014 IEEE International Solid-State Circuits Conference Digest of Technical Papers (ISSCC)* (Feb 2014), pp. 10–14
40. D. Ielmini, H.-S.P. Wong, In-memory computing with resistive switching devices. Nature Electronics **1**(6), 333–343 (2018)
41. H. Ghasemzadeh, N. Amini, R. Saeedi, M. Sarrafzadeh, Power-aware computing in wearable sensor networks: An optimal feature selection. IEEE Trans. Mobile Comput. **14**(4), 800–812 (2014)

42. S. Lauwereins, W. Meert, J. Gemmeke, M. Verhelst, Ultra-low-power voice-activity-detector through context and resource-cost-aware feature selection in decision trees, in *2014 IEEE International Workshop on Machine Learning for Signal Processing (MLSP)* (IEEE, 2014), pp. 1–6

43. K. Badami, S. Lauwereins, W. Meert, M. Verhelst, Context-aware hierarchical information-sensing in a 6μw 90nm CMOS voice activity detector, in *2015 IEEE International Solid-State Circuits Conference-(ISSCC)* (IEEE, 2015), pp. 1–3

44. J.-C. Pena-Ramos, K. Badami, S. Lauwereins, M. Verhelst, A fully configurable non-linear mixed-signal interface for multi-sensor analytics. IEEE J. Solid-State Circuits **53**(11), 3140–3149 (2018)

45. B. Murmann, M. Verhelst, Y. Manoli, Analog-to-information conversion, in *NANO-CHIPS 2030* (Springer, 2020), pp. 275–292

46. J.H. Korhonen, and P. Parviainen, Tractable Bayesian network structure learning with bounded vertex cover number, in *Advances in Neural Information Processing Systems* (2015), pp. 622–630

47. R. Livni, S. Shalev-Shwartz, O. Shamir, On the computational efficiency of training neural networks, in *Advances in Neural Information Processing Systems* (2014), pp. 855–863

48. Y. Huang, Y. Cheng, A. Bapna, O. Firat, D. Chen, M. Chen, H. Lee, J. Ngiam, Q.V. Le, Y. Wu, et al., GPipe: Efficient training of giant neural networks using pipeline parallelism, in *Advances in Neural Information Processing Systems* (2019), pp. 103–112

49. Q. Yang, Y. Liu, T. Chen, Y. Tong, Federated machine learning: Concept and applications. ACM Trans. Intell. Syst. Technol. (TIST) **10**(2), 1–19 (2019)

50. C. Elkan, The foundations of cost-sensitive learning, in *International Joint Conference on Artificial Intelligence*, vol. 17(1) (2001), pp. 973–978

51. Y. Wang, I.I. Hussein, D. Brown, R.S. Erwin, Cost-aware Bayesian sequential decision-making for search and classification. IEEE Trans. Aerosp. Electron. Syst. **48**(3), 2566–2581 (2012)

52. Z. Xu, M. Kusner, K. Weinberger, M. Chen, Cost-sensitive tree of classifiers, in *International Conference on Machine Learning* (2013), pp. 133–141

53. X. Chai, L. Deng, Q. Yang, C.X. Ling, Test-cost sensitive naive Bayes classification, in *Fourth IEEE International Conference on Data Mining (ICDM'04)* (IEEE, 2004), pp. 51–58

54. A. Verachtert, H. Blockeel, J. Davis, Dynamic early stopping for naive Bayes, in *Proceedings of the Twenty-Fifth International Joint Conference on Artificial Intelligence*, vol. 2016 (AAAI Press, 2016), pp. 2082–2088

55. Z. Xu, M.J. Kusner, K.Q. Weinberger, M. Chen, O. Chapelle, Classifier cascades and trees for minimizing feature evaluation cost. J. Mach. Learn. Res. **15**(1), 2113–2144 (2014)

56. H. Inoue, Adaptive ensemble prediction for deep neural networks based on confidence level, in *The 22nd International Conference on Artificial Intelligence and Statistics* (PMLR, 2019), pp. 1284–1293

# Chapter 2
# Background

This chapter lays out the relevant theoretical foundations. Section 2.1 reviews the principles of probability theory, introducing, first, some useful theorems and propositions. Then, it defines the concepts of probabilistic distributions and probabilistic inference. Finally, it discusses how independence and conditional independence notions have been leveraged to address some of the scalability challenges of probabilistic inference. The concepts discussed in this section are based on the books by Stuart Russell and Peter Norvig [1], David Poole and Alan Mackworth [2], and Adnan Darwiche [3].

Section 2.2 is dedicated to Bayesian networks, a type of model that can compactly encode conditional independence assertions in probabilistic models. This section introduces their semantics first. It then discusses the associated inference and learning techniques, as well as their use for classification and the task of focus throughout this book.

The work introduced in Chaps. 5 and 6 moves away from Bayesian networks and focuses on the exploitation of Probabilistic Circuits, a state-of-the-art representation that can guarantee tractable inference for a number of complex queries. Section 2.3 discusses their general semantic and syntactic properties, as well as how to use them for classification tasks.

Finally, Sect. 2.4 sets the stage for the range of tasks and applications considered in this book. In particular, this section narrows down the type of embedded devices the methods proposed in this book target. Furthermore, this section introduces the concept of the sensing embedded pipeline that will be used for the proposed methods throughout the chapters to follow.

L. I. Galindez Olascoaga et al., *Hardware-Aware Probabilistic Machine Learning Models*, https://doi.org/10.1007/978-3-030-74042-9_2

## 2.1  Probability Theory

Probability theory grew into one of the cornerstones of AI as it became clear that rational agents must act under inherently imprecise information about the world. This sub-field of mathematics models and quantifies uncertainty in terms of beliefs and studies how knowledge affects these beliefs [2]. A belief can be interpreted as the probability that a given proposition $\alpha$ is true. Thus if an agent's probability of $\alpha$ is greater than zero and less than one, it means that the agent is ignorant of whether $\alpha$ is true or false but holds a belief based on the knowledge it has access to. This knowledge could have been derived from statistical data, from expertise or from experience, or from a combination of sources.

### 2.1.1  Basic Notions and Notation

Probability theory is centered around the study of *random variables* that offer a representation of the possible outcomes in an experiment. The set of values a random variable can take on is its *domain*. *Discrete random variables* have a finite domain that can be expressed as a countable set. For example, the domain of a *Boolean variable* is the set {*true, false*}, while the domain of the roll of a die is the set of integers from 1 to 6.

This book uses the standard notation of probability theory: random variables are denoted by upper case letters $X$ and their instantiations by lower case letters $x$. If clear from the context, Boolean variable assignments are written as $x$ and $\neg x$ for $x = true$ and $x = false$, respectively. Sets of variables are denoted in bold upper case $\mathbf{X}$ and their joint instantiations in bold lower case $\mathbf{x}$. Finally, sets of variable sets are denoted with $\mathcal{X}$.

In probability theory it is common to refer to a world, denoted by $\omega$, as an assignment of values to the random variables under consideration. The set of all possible worlds is called the *sample space* and is denoted by $\Omega$. The roll of a pair of dice, for example, has a sample space of 36 possible worlds: (1,1),(1,2),....,(6,6). Thus, the total probability of the set of possible worlds must be equal to 1 [1]:

$$0 \leq P(\omega) \leq 1\, \forall \omega \quad \text{and} \quad \sum_{\omega \in \Omega} P(\omega) = 1. \tag{2.1}$$

In practical scenarios, one is usually interested in events that involve sets of possible worlds. The probability of an event $\psi$ is then determined by the sum of probabilities of worlds in which it holds

$$P(\psi) = \sum_{\omega \in \psi} P(\omega). \tag{2.2}$$

This formulation can help in answering queries that involve unconditional or prior knowledge. For example, "what is the probability that the roll of a pair of dice is equal to eleven?." In many practical scenarios, however, *evidence* is available and often comes from prior observations. In that case, one might be interested in the calculation of *conditional* or *posterior* probabilities. For the example with the pair of dice, one might want to know "what is the probability of doubles given that the first die was 5?." For any two propositions $a$ and $b$, the conditional probability of $a$ given $b$ is equal to

$$P(a|b) = \frac{P(a \wedge b)}{P(b)}.$$ (2.3)

## 2.1.2 Probability Distributions

A *probability distribution* defines the probabilities of the domain of a random variable. In the case of continuous variables, one can define it as parametrizable *probability density function*, such as a Gaussian or uniform distribution. However, in practice it is common to discretize continuous variables.

Distributions over multiple variables are called *joint probability distributions*. Table 2.1 shows an example of a three variable joint distribution for an activity recognition application, similar to the example in Fig. 1.5. Here, $R$ denotes the Boolean variable corresponding to the "Run" activity, $S$ corresponds to the standard deviation of an accelerometer signal in the vertical direction and is a Boolean variable taking values $\{low, high\}$,[1] and $M$ denotes the mean of an accelerometer signal in the horizontal direction, also with domain $\{low, high\}$. According to the axiom in Eq. 2.1, the elements in this tabular joint distribution must add up to 1. Moreover, Eq. 2.2 implies that the table can be used to calculate the probability of any proposition by adding the values of the cells that are consistent with it. For example, the probability of running ($P(r)$) can be calculated by adding the elements of the first row of the table: $P(r) = 0.011+0.151+0.093+0.347=0.602$. Extracting the probability over a subset of the available variables is a common calculation and is known as *marginalization* or *summing out* since the probabilities of the variables

**Table 2.1** Joint distribution over $R$ (Run), $S$ (Acc. vertical $\sigma$), and $M$ (Acc. horizontal $\mu$). Here, $s$ and $m$ denote a low value and $\neg s$ and $\neg m$ denote a high value for $S$ and $M$, respectively.

| | $m$ | | $\neg m$ | |
|---|---|---|---|---|
| | $s$ | $\neg s$ | $s$ | $\neg s$ |
| $r$ | 0.011 | 0.151 | 0.093 | 0.347 |
| $\neg r$ | 0.353 | 0.004 | 0.026 | 0.015 |

---

[1] For notation ease, the value *low* is denoted as $s$ and the value *high* is denoted as $\neg s$.

not included in the query are summed over. A general rule for marginalization is given by the following equation:

$$\Pr(\mathbf{Y}) = \sum_{\mathbf{z} \in \mathbf{Z}} \Pr(\mathbf{Y}, \mathbf{z}), \tag{2.4}$$

where all the combinations of values of the variable set $\mathbf{Z}$ are summed over.

### 2.1.3  Probabilistic Inference

*Probabilistic inference* is a methodology used to answer query propositions under observed evidence. One of the most common queries in practical scenarios is the calculation of conditional probability, where one is interested in the probability of a variable or sets of variables given the available evidence $\mathbf{e}$. Following the definition of conditional probability in Eq. 2.3, and taking into consideration the general rule for marginalization, the conditional probability can be calculated as follows:

$$\Pr(X|\mathbf{e}) = \frac{\Pr(X, \mathbf{e})}{\Pr(\mathbf{e})} = \beta \Pr(X, \mathbf{e}) = \beta \sum_{\mathbf{y}} \Pr(X, \mathbf{e}, \mathbf{y}), \tag{2.5}$$

where the denominator $\Pr(\mathbf{e})$ remains constant and can therefore be viewed as a normalization constant $\beta$. This notion is particularly useful for classification tasks when one must often compare the conditional probabilities of a variable given the same piece of observed evidence. One can then forego the calculation of $\Pr(\mathbf{e})$, which might not always be accessible or easy to compute. The $\mathbf{Y}$ variables in Eq. 2.5 remain unobserved and are therefore marginalized in the process of calculating $\Pr(X|\mathbf{e})$. Note that this equation assumes that $X$, $\mathbf{E}$, and $\mathbf{Y}$ constitute the complete set of variables and thus $\Pr(X, \mathbf{E}, \mathbf{Y})$ is the full joint distribution.

Equation 2.5 provides a general inference procedure for discrete variables. For example, one may want to know what are the probabilities of the values of $R$ given the observation that the accelerometer in the vertical direction has a low value ($s$):

$$\Pr(R|s) = \beta \Pr(R, s)$$
$$= \beta[\Pr(R, s, m) + \Pr(R, s, \neg m)]$$
$$= \beta[\langle \Pr(r, s, m), \Pr(\neg r, s, m) \rangle + \langle \Pr(r, s, \neg m), \Pr(\neg r, s, \neg m) \rangle]$$
$$= \beta[\langle 0.011, 0.353 \rangle + \langle 0.093, 0.026 \rangle] = \beta \langle 0.104, 0.379 \rangle = \langle 0.22, 0.78 \rangle,$$

where $\langle \cdot \rangle + \langle \cdot \rangle$ denotes vector sum. Thus, given evidence on a low value of $S$, the probability that the user is running is equal to 0.22, while the probability that the user is not running is 0.78. Note that, by including the notion of the normalization factor

$\beta$, there was no need to compute the denominator of the conditional probability formula in Eq. 2.3.

### 2.1.4  Independence Notions and Bayes Rule

It is clear that answering probabilistic queries with a full joint distribution, like in the example above, does not scale well for larger domains, as it requires an input table of size $\mathcal{O}(2^n)$ and takes time $\mathcal{O}(2^n)$. Most practical probabilistic inference approaches reduce this complexity by incorporating independence notions.

Suppose the example in Table 2.1 is extended with an additional variable $W$ (denoting the weather or general environmental conditions), with a domain of three values $\{fog, bright, cloudy\}$, denoted by $\{f, b, c\}$ for notation ease. The full joint distribution now has 24 entries (one eight-entry table for each type of condition). Suppose one has evidence over the activity recognition variables $R, S, M$ and wants to infer the probability of the weather given those variables.[2] Then, according to Eq. 2.3, the probability that it is cloudy ($W = c$) given that the user is running and the accelerometer readings are high ($e = \{r, \neg s, \neg m\}$) can be calculated with

$$\Pr(c|r, \neg s, \neg m) = \frac{\Pr(c, r, \neg s, \neg m)}{\Pr(r, \neg s, \neg m)}.$$

However, it is reasonable to assume that the probability of the weather being cloudy is the same regardless of the activity performed by the user and the wearable's sensor readings. The equation above can then be rewritten as

$$P(c|r, \neg s, \neg m) = P(c) = \frac{P(c, r, \neg s, \neg m)}{P(r, \neg s, \neg m)}.$$

That is, the probability of cloudy weather does not depend on any of the other instantiated variables. If this independence assertion is true for all involved variables and their values, the joint distribution can be rewritten as

$$\Pr(R, S, M, W) = \Pr(R, S, M)\Pr(W). \tag{2.6}$$

In general, independence between variables $X$ and $Y$ can be written as follows:

$$\Pr(X|Y) = \Pr(X), \quad \Pr(Y|X) = \Pr(Y), \tag{2.7}$$

$$\text{or} \quad \Pr(X, Y) = \Pr(X)\Pr(Y). \tag{2.8}$$

---

[2]Note that this example assumes that the possible environmental conditions do not influence the user's decision to perform a given activity.

These independence assertions imply that the full joint distribution can be *factored* into separate joint distributions. In the example above, the full distribution is factored into an 8 element distribution over $\{R, S, M\}$ and an independent 3 element distribution over $W$, which eliminates the need for the 24-entry table mentioned before.

However, often in practice, independence assertions like the one in the previous example are unlikely to occur. Besides, evidence is usually available over multiple variables. For activity recognition, for example, a relevant query could be *what is the probability that the user is running, given that the accelerometer in the vertical direction has a low standard deviation but the one in the horizontal direction has a high mean value?* This conditional probability is straightforward to compute from Table 2.1 but would not scale well for a scenario with more variables with larger domains. An alternative would be to reformulate the problem with Bayes rule, which states that:

$$\Pr(b|a) = \frac{\Pr(a|b)}{\Pr(a)} \cdot \Pr(b).\tag{2.9}$$

Thus, for the activity recognition example:

$$\Pr(R|s, \neg m) = \beta \cdot \Pr(s, \neg m|R) \cdot \Pr(R).$$

The problem with the evaluation of this equation is that the calculation of $\Pr(s, \neg m|R)$ might not scale up well, since there are $2^n$ combinations of observed values for which one would need to know the conditional probability. Moreover, variables $S$ and $M$ cannot be assumed to be independent since they are related by the activity the user is performing. Following the same reasoning, it could be assumed that high values for both accelerometer variables could be directly caused by the user's activity. Thus, if the user is indeed running, the two accelerometer variables can be assumed independent from each other because their values are directly explained by the activity. This assumption is known as *conditional independence* and allows to further reformulate the problem as follows:

$$\Pr(s, \neg m|R) = \Pr(s|R) \Pr(\neg m|R),$$

and therefore,

$$\Pr(R|s, \neg m) = \beta \Pr(s|R) \Pr(\neg m|R) \Pr(R).$$

Furthermore, conditional independence of $S$ and $M$ means that the joint probability distribution can be factored into three distributions:

$$\Pr(S, M, R) = \Pr(S, M|R) \Pr(R)$$
$$= \Pr(S|R) \Pr(M|R) \Pr(R).$$

The size of the original tabular representation grows with complexity $\mathcal{O}(2^n)$, whereas the representation above grows with $\mathcal{O}(n)$. This is because every table in this factored distribution has at most two independent parameters, whereas the original table can have up to seven.

Conditional independence assertions thus allow probabilistic models to scale well to scenarios with larger domains and more worlds. They are also easier to identify and model in practice.

To generalize the example above, the definition of conditional independence for two variables $X, Y$ given a third $Z$ is given by

$$\Pr(X, Y | Z) = \Pr(X | Z) \Pr(Y | Z). \tag{2.10}$$

Thus, the joint distribution over the three variables can be formulated as

$$\Pr(X, Y, Z) = \Pr(X | Z) \Pr(Y | Z) \Pr(Z). \tag{2.11}$$

## 2.2 Bayesian Networks

*Bayesian networks*, proposed by Judea Pearl [4], are a type of probabilistic model that offers a systematic way of encoding the conditional independence relationships discussed above. Bayesian networks can represent any joint probability distribution compactly through two main components. The first one is a Directed Acyclic Graph (DAG), where each node corresponds to a random variable and the presence or absence of edges among them specifies the conditional independence relations that hold in the modeled situation. An arrow from node $Y$ to $X$ indicates that $Y$ is the *parent* of $X$ and that $X$ is the *child* of $Y$. The second component follows from this structure: there is a conditional probability distribution associated to each node $X_i$, given by $\Pr(X_i | Parents(X_i))$, and for discrete variables, it can be expressed in a tabular format called a Conditional Probability Table (CPT), as shown in Fig. 2.1.

**Fig. 2.1** Bayesian network for an activity recognition and tracking application, based on the classical "Burglary" example by Judea Pearl [1]

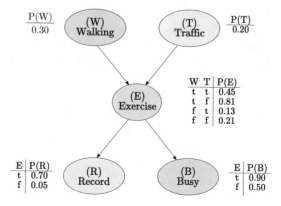

Bayesian networks represent joint probability distributions over the set of variables $\mathbf{X}$:

$$\Pr(X_1, \ldots, X_n) = \prod_{i=1}^{n} \Pr(X_i | Parents(X_i)). \tag{2.12}$$

Figure 2.1 shows an example of a Bayesian network corresponding to activity recognition and tracking in a smartphone. In this example, the user's smartphone can automatically identify whether the current activity is a form of exercise $(E)$. For example, a user may go for a walk $(W)$ either to exercise or run errands. Depending on the user's habits, background noise—such as that produced by traffic $(T)$—could also help in identifying whether the user is exercising. This Bayesian network also represents users' preferences on recording exercise-related activities $(R)$ and setting the status of their business communication apps to "busy" $(B)$ when exercising. The joint distribution in this example is given by

$$\Pr(W, T, E, R, B) = \Pr(W) \Pr(T) \Pr(E|W, T) \Pr(R|E) \Pr(B|E).$$

This representation of a joint distribution allows the straightforward calculation of several queries. For example, the probability that a user walks in order to run errands instead of exercising $(w, \neg e)$, in the absence of traffic noise $(\neg t)$, and desires to record this activity $(r)$ but not to appear busy in business communication platforms $(\neg b)$ can be calculated as follows:

$$P(w, \neg t, \neg e, r, \neg b) = P(w) P(\neg t) P(\neg e|w, \neg t) P(r|\neg e) P(\neg b|\neg e)$$

$$= 0.30 \cdot 0.80 \cdot 0.19 \cdot 0.05 \cdot 0.50 = 0.00114.$$

This ability to model and represent conditional independence assertions makes Bayesian networks a significantly more compact representation than the full joint distribution. The level of compactness of Bayesian networks is clearly determined by the ordering of the nodes (in terms of *parent–child* structure). In general, when determining the structure, it is a good practice to utilize expert knowledge and exploit causal relations among variables.[3] In Fig. 2.1, a non-clever ordering would, for example, be *Record* $\rightarrow$ *Busy* $\rightarrow$ *Walking* $\rightarrow$ *Traffic* $\rightarrow$ *Exercise* because it cannot exploit the conditional independence existing between the decision to record the activity and the decision to show a busy status, given the identification of an exercise activity. Instead, the example follows a more intuitive "causal" ordering where walking or traffic noise might lead to identifying exercising, and that might motivate the user to record the activity or to set a new status.

---

[3]Note that, according to Pearl's and Mackenzie's recent book [5], Bayesian networks don't strictly encode causal relations, they are statistical models.

There are two topological criteria in Bayesian networks that specify conditional independence relations [1, 4]. The first one is that "each variable is conditionally independent of its non-descendants, given its parents" [1]. For example, in Fig. 2.1 the user's decision to record the activity is prompted by exercising and not by the status of the business communication app. The second property encoded by the DAG is that "a node is conditionally independent of all other nodes in the network, given its parents, children, and children's parents" [1]. These properties are useful in determining what structure is suitable for the desired complexity level of the model in the application of interest, and also to readily identify the independence relations among variables when the structure of the Bayesian network is provided.

## 2.2.1 Exact Inference in Bayesian Networks

Recall from Eq. 2.5 that conditional probability queries can be answered by summing out unobserved variables from the joint distribution. Bayesian networks allow to express the full joint distribution as a product of conditional distributions. Thus, inference in Bayesian networks takes place with a series of multiplications and additions.

Consider the example in Fig. 2.1 and suppose one of the smartphone's apps needs to know the probability that the user is walking ($w$). The app will likely not query the user to gain this information. Instead, it needs to infer this probability from the state of the activity recording app (e.g. currently not recording, or $\neg r$) and the user's status on the business communication app (e.g. set to busy, or $b$). This situation constitutes a conditional probability query, as defined in Eq. 2.5 and calculated as follows for the current example:

$$\Pr(w|\neg r, b) = \beta \sum_t \sum_e P(w)P(t)P(e|w, t)P(\neg r|e)P(b|e).$$

This calculation can be performed by adding four terms (for each possible combination of $T$ and $E$), each obtained by multiplying five numbers. In the worst case, when having to marginalize most (Boolean) variables, the complexity of this sum–product algorithm can be $\mathcal{O}(n2^n)$. The variable elimination algorithm can reduce the complexity substantially by following a dynamic programming approach that avoids the repetition of computations that have already been performed.

For the example network, the following expression can be evaluated, where each probability distribution is a *factor* and is denoted by $\phi$:

$$\Pr(W|\neg r, b) = \beta \underbrace{\Pr(W)}_{\phi_1(W)} \sum_t \underbrace{P(t)}_{\phi_2(T)} \sum_e \underbrace{\Pr(e|W, t)}_{\phi_3(E,W,T)} \underbrace{P(\neg r|e)}_{\phi_4(E)} \underbrace{P(b|e)}_{\phi_5(E)}.$$

The process iteratively sums out factors from right to left. The first step is to sum out $E$ from the products of $\phi_3$, $\phi_4$, and $\phi_5$ and save it into a new factor $\phi_6$:

$$\Pr(W|\neg r, b) = \beta\phi_1(W) \cdot \sum_t \phi_2(T) \cdot \phi_6(W, T). \tag{2.13}$$

The following step would be to sum out $T$, and so on. The complexity of the variable elimination algorithm is determined by the largest intermediate factor that in turn depends on the order in which variables are summed out and factors are multiplied.

In general, the complexity of exact inference depends strongly on the network's structure. The time and space complexity of inference on singly connected networks (or poly-trees) like the example above is linear in the size of the network. But for multiple connected networks, where there is more than one path between any two nodes, variable elimination can have exponential time and space complexity.

### 2.2.2 Parameter Learning

Given their structure—determined, for example, from expert knowledge—the parameters in the CPTs of Bayesian networks can be estimated from data, by a process called maximum likelihood estimation, as explained in [3].

This estimation approach is based on the assumption that the available dataset $\mathcal{D}$ of size $N$ was simulated from the Bayesian network that models the *true* distribution of the random variables in the domain of interest. Under this assumption the empirical distribution of an instantiation $\gamma$ is given by

$$\Pr_{\mathcal{D}}(\gamma) = \frac{\mathcal{D}\#(\gamma)}{N}, \tag{2.14}$$

where $\mathcal{D}\#(\gamma)$ is the number of cases in the dataset satisfying $\gamma$. The maximum likelihood estimate of a CPT parameter $\hat{\theta}_{x|parents(x)}$ can be calculated from this empirical distribution:

$$\hat{\theta}_{x|parents(x)} = \Pr_{\mathcal{D}}(x|parents(x)) = \frac{\mathcal{D}\#(x, parents(x))}{\mathcal{D}\#(parents(x))}. \tag{2.15}$$

This parameter estimate maximizes the likelihood function [3]:

$$\theta^* = \mathrm{argmax}_\theta L(\theta|\mathcal{D}), \tag{2.16}$$

where $L(\theta|\mathcal{D})$ is given by

$$L(\theta|\mathcal{D}) = \prod_{i=1}^{N} \Pr(\mathbf{d}_i|\theta). \tag{2.17}$$

Here, $\Pr(\cdot|\theta)$ is the probability distribution induced by a Bayesian network parametrized with estimates $\theta$, and $\mathbf{d}_i$ is called a *case* of dataset $\mathcal{D}$ and is a vector with values of variables $\mathbf{X}$.

### 2.2.3 Bayesian Network Classifiers

The task of classification consists in assigning a *class label* $C$ to instances described by a set of *features* $\mathbf{F}$. This task is done by applying Bayes rule (Eq. 2.9) to compute the probability of a class $C$ given a particular instance $\mathbf{f}$ and predicting the class value $c$ that maximizes the posterior probability:

$$c = \underset{C}{\mathrm{argmax}}\, \Pr(C|\mathbf{f}). \qquad (2.18)$$

Despite its simplicity and strong conditional independence assumptions, naive Bayes classifiers are competitive in terms of classification performance. However, some of the independence assumptions they encode are often not supported by the data. *Bayesian network classifiers* were introduced in [6] to improve the performance of naive Bayes by relaxing some of their strong independence biases.

Bayesian network classifiers based on naive Bayes structure[4] require the class variable to be a parent to every feature. The term $\Pr(C|\mathbf{F})$ is always encoded in the network, ensuring that all features depend on the class variable. The left side of Fig. 2.2 shows an example of a naive Bayesian network classifier that models a scenario where a mobile robot has to identify what room it is in based on some of its on-board measurements and observations.

The performance of Bayesian network classifiers can be further improved with respect to the naive structure by augmenting it with edges among the features and

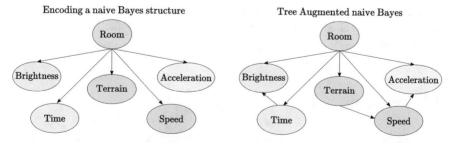

**Fig. 2.2** Bayesian network classifiers for a robot localization scenario. The goal for the robot is to identify what room it is in based on its speed, acceleration, as well as the roughness of the terrain, the time of the day, and the level of environmental brightness

---

[4]These are the type of Bayesian network classifiers used throughout this book.

thus encoding the relations between features that are found in many domains. Finding the optimal set of augmenting edges among features is an intractable problem, so Friedman et al. proposed a restricted structure called the Tree Augmented naive Bayes (TAN) classifier where each feature has as parents the class variable and at most one other feature [6]. The right side of Fig. 2.2 shows an example of a TAN classifier, where the acceleration of the robot is influenced by its speed, and its speed by the type of terrain it navigates in.

The procedure for learning the edges of a TAN classifier from data has polynomial time and is based on Chow and Liu's method for tree learning [7], which builds a maximum weighted spanning tree based on the mutual information between pairs of features, and has been proven to find the tree that maximizes the likelihood of the model given the data. The general definition of mutual information between pairs of variables $\mathbf{X}$ and $\mathbf{Y}$ is

$$\mathrm{MI}(\mathbf{X}, \mathbf{Y}) = \sum_{\mathbf{x},\mathbf{y}} P(\mathbf{x}, \mathbf{y}) \log \frac{P(\mathbf{x}, \mathbf{y})}{P(\mathbf{x})P(\mathbf{y})}. \tag{2.19}$$

The method for learning TAN classifiers constructs the maximum weighted spanning tree by considering the conditional mutual information of pairs of features given the class variable. Conditional mutual information of variables $\mathbf{X},\mathbf{Y}$ given $\mathbf{Z}$ is given by

$$\mathrm{CMI}(\mathbf{X}, \mathbf{Y}|\mathbf{Z}) = \sum_{\mathbf{x},\mathbf{y},\mathbf{z}} P(\mathbf{x}, \mathbf{y}, \mathbf{z}) \log \frac{P(\mathbf{x}, \mathbf{y}|, \mathbf{z})}{P(\mathbf{x}|\mathbf{z})P(\mathbf{y}|, \mathbf{z})}. \tag{2.20}$$

The inference methods discussed in Sect. 2.2.1 can be used to compute the class posterior probability in Eq. 2.18. Naturally, the time and space complexity of the naive Bayes structure is lower than that of the TAN classifier.

There is a clear trade-off between the expressiveness of a model and its efficiency, in which the efforts described in the following section have striven to balance, resulting in a new class of probabilistic models that guarantee tractability for several complex queries.

## 2.3  Probabilistic Circuits

As discussed throughout the last section, exact probabilistic inference can have exponential complexity for certain models and queries. Approximate inference offers an alternative to deal with restrictive computational bandwidth but does not come with performance guarantees and can still remain intractable [8, 9].

This has given rise to the field of tractable probabilistic modeling, which concerns methods that aim to balance the computational complexity versus expressiveness trade-off of probabilistic inference. Formally, a query $q(m)$ is tractable on model $m$ if and only if it can exactly be computed in time $\mathcal{O}(poly(|m|))$ [10].

The term probabilistic circuit (PC) was recently introduced in [10] as a general computational framework that unifies the variety of formalisms and representations targeting tractable probabilistic inference, such as arithmetic circuits [11], Cutset Networks [12], Sum–Product Networks [13], and Probabilistic Sentential Decision Diagrams [14].

As a general framework, PCs are computational graphs that encode joint probability distributions in a recursive way. Even though they are defined in terms of a graphical formalism, they are not a type of Probabilistic Graphical Model (PGM) (like Bayesian networks). In fact, the nodes in a PC directly represent how to evaluate a probability (e.g. a marginal). That is, they have clearly defined *operational semantics*. By constraining their graph structure, PCs are capable of tractable inference while remaining expressive. Furthermore, they allow to encode more fine grained relations among variables and values than Bayesian networks, which mostly rely on conditional independence, thus allowing for further factorization of the joint distribution they represent.

### 2.3.1 Properties of Probabilistic Circuits

A PC over random variables $\mathbf{X}$ is characterized by a graphical structure $\mathcal{G}$ and a set of parameters $\boldsymbol{\theta}$.

**Structure and Parameters** The nodes in a PC are computational units of two types: (1) internal nodes that alternate between multiplications and additions and (2) input (or leaf) nodes that encode univariate distributions over each random variable $X_i \in \mathbf{X}$. Common choices for these distributions are exponential families, such as Gaussians, and Bernoulli for Boolean variables. In contrast to PGMs, edges between the nodes of a PC denote the order of execution of the computational graph that is evaluated form bottom to top. The set of parameters of a PC is $\boldsymbol{\theta} = \boldsymbol{\theta_S} \cup \boldsymbol{\theta_L}$, where $\boldsymbol{\theta_S}$ weighs the edges of sum nodes (as shown by $\theta_S$ in the example of Fig. 2.3) and $\boldsymbol{\theta_L}$ parametrizes the leaf distributions (as denoted by $\theta_A$ and $\theta_B$ in Fig. 2.3).

**Computational Semantics** Each node $n$ in a PC is defined over a subset of $\mathbf{X}$, as determined by a scope function $\phi$, i.e. $\phi(n) \subseteq \mathbf{X}$. This function is defined recursively: for each leaf, the scope is the variable $X_i \in \mathbf{X}$ over which its distribution is defined; the scope of an internal node $n$ is given by $\phi(n) = \cup_{c \in children(n)} \phi(c)$; and the scope of the root node is the full variable set $\mathbf{X}$. Product nodes define the factorization $q_n(\mathbf{X}) = \prod_{c \in children(n)} q_c(\mathbf{X})$, while sum nodes define a weighted sum $q_n(\mathbf{X}) = \sum_{c \in children(n)} \theta_{S,c} q_c(\mathbf{X})$ with $\sum_c \theta_{S,c} = 1$.

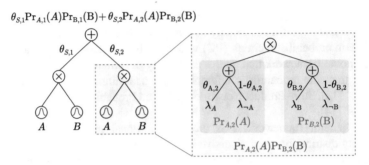

**Fig. 2.3** Example of a Probabilistic Circuit with input Bernoulli distributions over Boolean variables $\{A, B\}$

Figure 2.3 shows an example of a PC over Boolean[5] variables $A$ and $B$. This PC encodes the distribution $\mathrm{Pr_{out}}(A, B) = \theta_{S,1} \mathrm{Pr}_{A,1}(A) \mathrm{Pr}_{B,1}(B) + \theta_{S,2} \mathrm{Pr}_{A,2}(A) \mathrm{Pr}_{B,2}(B)$, where, e.g. $\mathrm{Pr}_{A,1}(A)$ is a Bernoulli distribution over $A$ parametrized by $\theta_{A,1}$. Given the evidence, the probability at the output of this PC can be evaluated by setting the Boolean indicator variables $\lambda$ accordingly. For example, if $\mathbf{e} = \{A = \text{true}, B = \text{false}\}$, the indicators should be set to $\lambda_A = 1$, $\lambda_{\neg A} = 0$, $\lambda_B = 0$, $\lambda_{\neg B} = 1$. The PC can also be evaluated under incomplete evidence, e.g. $\mathbf{e} = \{B = \text{false}\}$. In this case variable $A$ can be marginalized by setting both $\lambda_A$ and $\lambda_{\neg A}$ to 1 and setting the other indicators accordingly.

The work discussed in this book focuses on classification tasks that rely on conditional queries. Equation 2.3 will be used to evaluate conditional probability. Therefore, only two types of queries are considered: queries with complete evidence, and marginal queries.

### 2.3.2 Structural Constraints

The computation of marginal probabilistic queries in PCs is guaranteed to be tractable when they meet the two following structural constraints [15]:

**Decomposability** Decomposability: A PC is decomposable if the children of all its product nodes are disjoint. For example, in Fig. 2.3, the left child of the product nodes is defined over $A$, whereas the right child is defined over $B$.

**Smoothness** Smoothness: A PC is smooth if the children of all its sum nodes depend on the same variable sets. For example, both the left and right children of the sum node in Fig. 2.3 are defined over $\{A, B\}$.

---

[5]The work introduced in Chaps. 5 and 6 focuses on Boolean variables with Bernoulli distributions. A one-hot encoding is used to deal with multi-valued random variables under this restriction.

A third important structural constraint found in some PCs is determinism, which enables other tractable queries such as Maximum a Posteriori (MAP) inference. The discussion herein is limited to marginal queries, and refer the reader to the literature for a more elaborate overview.

**Determinism**  Determinism: A sum node is deterministic if the value of at most one of its children (or inputs) is nonzero. A PC is deterministic if all its sum nodes are deterministic.

The work introduced in Chaps. 5 and 6 relies on the Probabilistic Sentential Decision Diagram (PSDD), whose properties are discussed throughout Sect. 5.1.1. In particular, those chapters leverage the LEARNPSDD algorithm [16], which learns the structure of the model as well as its parameters iteratively and directly from data.

Another more general type of PC considered in this book is the Arithmetic Circuit (AC), first introduced in [11]. In particular, ACs will be used for benchmarking: in Chap. 5 the accuracy and cost of Bayesian network classifiers are compared to the proposed method by transforming them to ACs via a process known as *knowledge compilation* [3]. In this process, the local structure of a PGM is encoded as a propositional logic formula and it is then compiled to another more succinct target logical form. Inference can then be performed by *weighted model counting* [17].

## 2.3.3   Classification Tasks with Probabilistic Circuits

Probabilistic Circuits allow to compute marginal probability queries. Similar to naive Bayes and Bayesian networks, classification can be performed with PCs by computing Bayes rule:

$$\Pr(C|\mathbf{F}) = \frac{\Pr(\mathbf{F}|C)\Pr(C)}{\Pr(\mathbf{F})} = \frac{\Pr(C,\mathbf{F})}{\Pr(\mathbf{F})}, \qquad (2.21)$$

and selecting the class value that maximizes it.

Probabilistic models are selected to fulfill the machine learning tasks discussed in this book because they are amenable to resource-constrained portable applications, which must remain robust under dynamically changing conditions. The efficient implementation of probabilistic models is possible by proposing a series of techniques that exploit the scalability of multiple layers of the device hosting the machine learning algorithm. The following section provides an overview of such devices' properties, identifying the trade-offs leveraged by the hardware-aware techniques proposed throughout this book. The stages executed by the device, from the moment it gathers sensory data to the execution of the machine learning task, are referred to as the *sensory embedded pipeline*. The hardware characterization of the following chapter will rely on this concept.

## 2.4 Sensory Embedded Pipeline

Figure 2.4 illustrates the sensory embedded pipeline that the devices considered in this book execute. The goal of this pipeline is to extract knowledge from the sensory observations made locally and perform a task of interest as efficiently as possible. An example of such a pipeline for a portable activity recognition application was shown in Fig. 1.6 and some of its implications were discussed in Sect. 1.3.4. Recall from that discussion that one of the remaining challenges towards extreme-edge intelligence is the fact that each stage in the pipeline has a specific contribution towards the overall resource consumption of the system,[6] yet some of these contributions are sometimes ignored or considered in isolation by current resource-aware approaches. Thus, one of the goals of studying the embedded sensing pipeline is to map the resource consumption contributions of each of the stages towards a common, system-wide cost metric, given in terms of energy consumption. Furthermore, these mappings allow to identify cost-saving opportunities at each stage procured by scalability opportunities, such as noise tolerance or computational precision tuning. Section 2.4.1 provides an overview of the main blocks in the pipeline.

### 2.4.1 Building Blocks

The sensory embedded pipeline consists of the three following building blocks, as illustrated in Fig. 2.4:

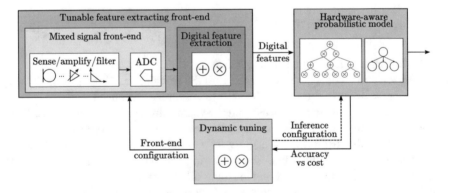

**Fig. 2.4** Sensory embedded pipeline

---

[6]Recall that this book focuses on energy consumption as the main proof of concept.

**Tunable Feature Extracting Front-End** The incoming sensory signal is first processed by a set of analog circuits—among which are amplifiers and filters—whose tolerance to noise can be scaled. The processed signal is then discretized with an Analog to Digital Converter (ADC) and sent out to the digital processing sub-block for the extraction of features. These digital features can be effectively employed by the algorithms within the two other blocks. Features are typically extracted in the digital domain, after the processed signal has been discretized by the ADC. However, some front-ends are capable of extracting features in the analog domain, aiming to minimize the volume of data to be processed in the digital domain [18]. Sections 3.2.1 and 3.3 will describe the implications in terms of feature quality and resource consumption of this block. Note also that each sensory signal is interfaced to its own processing chain and can therefore be individually tuned, a property that will be exploited by the frameworks introduced in Chap. 4 and illustrated by the activity recognition example of Fig. 1.6.

**Hardware-Aware Probabilistic Model** The hardware-aware probabilistic model performs the machine learning task. This model is capable of exploiting different configurations of the embedded pipeline when fulfilling the task it is trained for. For example, it can classify incoming samples by extracting one feature with lower quality and performing inference with less bits. A non-hardware-aware probabilistic model may experience performance degradation when operating under such a configuration. Two instances of this model are proposed in Chaps. 4 and 5, one of them is a Bayesian Network classifier, and the other one is a Probabilistic Circuit.

**Dynamic Tuning Block** Finally, the dynamic tuning block allows to tune, at run-time, the configuration of the feature extracting front-end. Strategies utilizing such a block and targeting the hardware-aware models mentioned above are proposed in Chap. 6.

Chapter 3 proposes the *hardware-aware cost*, a metric that abstracts the resource consumption resulting from the scalability enabled by the device. This metric constitutes the basis of the hardware-aware optimization strategies proposed throughout the rest of this book.

# References

1. S. Russel, P. Norvig, et al., *Artificial Intelligence: A Modern Approach* (Pearson Education Limited, 2013)
2. D.L. Poole, A.K. Mackworth, *Artificial Intelligence: Foundations of Computational Agents* (Cambridge University Press, 2010)
3. A. Darwiche, *Modeling and Reasoning with Bayesian Networks* (Cambridge University Press, 2009)
4. J. Pearl, *Probabilistic Reasoning in Intelligent Systems: Networks of Plausible Inference* (Morgan Kaufmann, 1988)
5. J. Pearl, D. Mackenzie, *The Book of Why: The New Science of Cause and Effect* (Basic Books, 2018)

6. N. Friedman, D. Geiger, M. Goldszmidt, Bayesian network classifiers. J. Mach. Learn. **29**(2), 131–163 (1997)
7. C. Chow, C. Liu, Approximating discrete probability distributions with dependence trees. IEEE Trans. Inf. Theory **14**(3), 462–467 (1968)
8. P. Dagum, M. Luby, Approximating probabilistic inference in Bayesian belief networks is NP-hard. Artificial Intelligence **60**(1), 141–153 (1993)
9. D. Roth, On the hardness of approximate reasoning. Artificial Intelligence **82**(1–2), 273–302 (1996)
10. Y. Choi, A. Vergari, G. Van den Broeck, Lecture notes: Probabilistic circuits: Representation and inference (2020). [Online]. Available: http://starai.cs.ucla.edu/papers/LecNoAAAI20.pdf
11. A. Darwiche, A differential approach to inference in Bayesian networks. J. ACM (JACM) **50**(3), 280–305 (2003)
12. T. Rahman, P. Kothalkar, V. Gogate, Cutset networks: A simple, tractable, and scalable approach for improving the accuracy of Chow-Liu trees, in *Joint European Conference on Machine Learning and Knowledge Discovery in Databases* (Springer, 2014), pp. 630–645
13. H. Poon, P. Domingos, Sum-product networks: A new deep architecture, in *2011 IEEE International Conference on Computer Vision Workshops (ICCV Workshops)* (IEEE, 2011), pp. 689–690
14. D. Kisa, G. Van den Broeck, A. Choi, A. Darwiche, Probabilistic sentential decision diagrams, in *Fourteenth International Conference on the Principles of Knowledge Representation and Reasoning* (2014)
15. A. Darwiche, P. Marquis, A knowledge compilation map. J. Artif. Intell. Res. **17**, 229–264 (2002)
16. Y. Liang, J. Bekker, G. Van den Broeck, Learning the structure of probabilistic sentential decision diagrams, in *Proceedings of the Conference on Uncertainty in Artificial Intelligence (UAI)* (2017)
17. M. Chavira, A. Darwiche, On probabilistic inference by weighted model counting. Artificial Intelligence **172**(6–7), 772–799 (2008)
18. K.M. Badami, S. Lauwereins, W. Meert, M. Verhelst, A 90 nm CMOS, power-proportional acoustic sensing frontend for voice activity detection. IEEE J. Solid State Circuits **51**(1), 291–302 (2016)

# Chapter 3
# Hardware-Aware Cost Models

One of the most restrictive aspects of extreme-edge IoT nodes is that they tend to be battery-powered in an attempt to seamlessly integrate them to the physical world, as discussed in Sect. 1.1.1. Energy consumption management is therefore crucial to their always-on functionality, especially when they must execute complex algorithms or signal processing. The proofs of concept in this book therefore focus on the trade-off between energy consumption and classification accuracy. This choice is also motivated by hardware innovations that pursue energy-efficient device operation [1, 2]. The following chapters build upon some of the theoretical properties of these systems, such as the effect that the choice of signal quality has on energy consumption, as derived in Sect. 3.2.1, or the fact that energy consumption can scale with digital precision, as discussed in Sect. 3.3.

The overarching goal of this book is to systematically trade-off energy consumption savings for limited classification accuracy losses. This implies that performance will be evaluated in terms of the relative energy consumption savings, and resulting accuracy losses, and that these relative trends will be incorporated into the trade-off optimization strategies. The contributions of this book therefore rely on the concept of *hardware-aware cost*, a system-wide energy consumption metric that expresses the scalability available in the device. This metric is modular, since it integrates the different contributions of the embedded sensing pipeline in an additive way. The following chapters of this book can therefore exploit the model of the particular stages of interest in the pipeline.

In particular, the methods and models proposed in Chap. 4 will focus on systems where the energy consumption of the sensor front-end dominates, while the ones in Chap. 5 focus on systems with more advanced digital processing, where the contributions from both the sensor front-end and the block implementing the machine task must be factored in.

L. I. Galindez Olascoaga et al., *Hardware-Aware Probabilistic Machine Learning Models*,
https://doi.org/10.1007/978-3-030-74042-9_3

Section 3.1 first establishes the general notion of hardware-aware cost. Then, Sects. 3.2, 3.3, and 3.4 discuss how energy scales in the front-end and inference blocks and define the costs of these blocks accordingly. Finally, Sect. 3.5 establishes the costs incurred by dynamic tuning strategies. The types of systems considered throughout this book are identified by Sect. 3.6, and Sect. 3.7 closes with concluding remarks.

## 3.1  Hardware-Aware Cost

This section provides a general definition of the hardware-aware cost, which represents the system-wide energy consumption of the sensory embedded pipeline introduced in Sect. 2.4. The hardware-aware cost considers the contributions associated with sensing and signal processing, digital feature extraction and inference, and run-time tuning. This metric can be customized to represent properties of the target scalable hardware, as will be discussed throughout this chapter.

Let $\mathcal{M}$ be a probabilistic model that encodes a joint probability distribution over variables $\mathbf{F}$, extracted from the set of sensor interfaces $\mathbf{S}$. The hardware-aware cost is defined as (see Fig. 3.1):

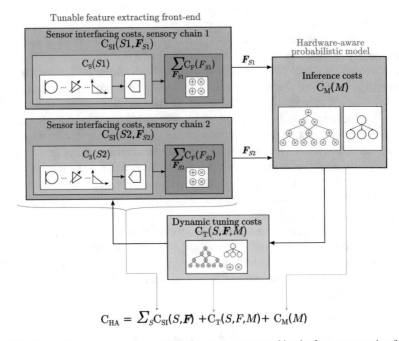

$$C_{HA} = \sum_S C_{SI}(S, \mathbf{F}) + C_T(S, F, M) + C_M(M)$$

**Fig. 3.1** The hardware-aware cost considers the energy consumed by the feature extracting front-end, by the probabilistic model and by the dynamic tuning block when deployed

$$C_{HA}(\mathcal{M}, \mathbf{S}, \mathbf{F}) = \sum_{S \in \mathbf{S}} C_{SI}(S, \mathbf{F}_S) + C_M(\mathcal{M}) + C_T(\mathcal{M}, \mathbf{S}, \mathbf{F}), \qquad (3.1)$$

where $C_{SI}$ are the sensor interfacing and feature extraction costs, with $\mathbf{F}_S$ the feature subset extracted from sensor $S$; $C_M$ the inference costs, pertaining to the execution of the machine learning task on model $\mathcal{M}$; and $C_T$ the costs corresponding to the dynamic tuning block, deployed for run-time strategies. Sensor interfacing costs are, in turn, defined as

$$C_{SI}(S, \mathbf{F}_S) = C_S(S) + \sum_{F_S \in \mathbf{F}_S} C_F(F_S), \qquad (3.2)$$

where $C_S$ describes the cost incurred by sensor $S$ and its mixed-signal front-end, and $C_F$ the cost of extracting feature set $\mathbf{F}_S \subseteq \mathbf{F}$ in the digital domain. Note that, if no features from a given sensor are used, it can be shut down, and its cost $C_S$ dropped. Similarly, the cost of the dynamic tuning strategy $C_T$ will only be accounted for when deploying it at run-time.

The rest of this chapter defines each term in Eq. (3.1) based on its hardware properties and its scaling capabilities. Specifically, Sect. 3.2 discusses the costs incurred by the device's sensors and their mixed-signal front-ends, based on their physical properties. Then, Sect. 3.3 specifies the costs associated with extracting features from the processed sensory signal. Finally, Sects. 3.4 and 3.5 define the costs associated with inference in the probabilistic model and with the execution of run-time tuning strategies, respectively.

## 3.2   Sensing Costs

The sensing cost $C_S$ is determined by the type of sensor used, and by the properties of its front-end. Recall from Sect. 2.4.1 that the incoming sensory signal is first processed by a set of analog circuits, among which are amplifiers and filters, whose tolerance to noise can be scaled. This signal is then discretized with an ADC for subsequent processing in the digital domain.

The sensor front-end can be tuned to meet the desired signal quality, which, in turn, determines its energy consumption. The following section derives the relation between signal quality and energy consumption, on which the definition of the sensing cost $C_S$ is based.

**Fig. 3.2** Mixed-signal sensor front-end. The sensor is modeled as an input voltage and the analog amplification and filtering as an inverting amplifier with a capacitive feedback network

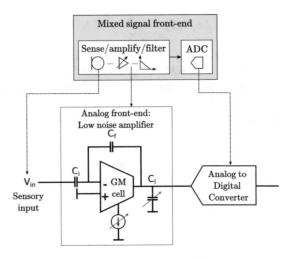

### 3.2.1   Resource Versus Quality Trade-Offs of Mixed-Signal Sensor Front-Ends

This section derives the models that establish the trade-off between the quality of the processed sensory signal within the front-end (in terms of how much noise is tolerated) and energy consumption.[1]

Figure 3.2 shows the properties of the mixed-signal front-end: the sensor is modeled as input voltage $V_{in}$, which is then amplified by an analog low-noise amplifier, and finally digitized by an ADC for further use by digital domain blocks.

This book considers two main sources of noise arising at the sensor front-end: the *circuit noise* of the amplifier and the *quantization noise* of the ADC.

The amplification stage within the sensor front-end can be modeled as an inverting amplifier with a capacitive feedback network, and the ADC as an *nb* bit quantizer, operating at a sampling frequency of $f_{samp}$. The amplifier itself is based on a transconductance ($Gm$) cell, whose differential input voltage produces an output current, while the feedback network determines the gain of the amplifier. The mean-square value of the voltage noise at the output of the $Gm$ cell is given by

$$V_{n,gm,out}^2 = \left(1 + \frac{C_i}{C_f}\right)\frac{2\alpha\gamma kT}{C_L}, \tag{3.3}$$

where $\alpha = 1 + gm_{bias}/gm_{inp}$ and is typically between 1 and 2; $\gamma$ is a process dependent constant typically between 2.3 and 2.5, $k$ the Boltzmann-constant and $T$ the operating temperature. Amplifier noise is predominantly determined by the

---

[1]These model derivations were first introduced in [3] and are the result of a collaboration with Komail Badami, who focused on the analog circuits aspects.

transconductance ratio $gm_{bias}/gm_{inp}$, which must be defined to meet the desired noise specifications.

The quantization noise from the ADC completing the front-end is given by [4]

$$V^2_{n,quant,ADC} = \frac{1}{12} \cdot \frac{V^2_{ref}}{2^{nb}}, \tag{3.4}$$

where $V_{ref}$ is the ADC reference voltage.

Assuming that the two noise sources are uncorrelated, the total noise contribution of the sensing front-end, $V^2_{n,rms,tot}$ can be expressed as

$$V^2_{n,rms,tot} = V^2_{n,gm,out} + V^2_{n,quant,ADC}. \tag{3.5}$$

For a normalized signal swing of $1 V_{p-p}$, the signal quality at the front-end output can be expressed in terms of Signal to Noise Ratio (SNR) as

$$SNR = -20 \log(2\sqrt{2}V_{n,rms,tot}), \tag{3.6}$$

or, in terms of *effective number of bits (enob)*:

$$enob = \frac{SNR - 1.76}{6.02}. \tag{3.7}$$

This notion compares the relative quality of a signal with respect to noise.

Analogous to sensor front-end noise, energy consumption of the front-end is determined by the individual contributions of the amplifier and the ADC:

$$E_{tot} = E_{gm,out} + E_{ADC}. \tag{3.8}$$

The energy consumption of the ADC is given by

$$E_{ADC} = 2^{enob} FOM_{ADC}, \tag{3.9}$$

where $FOM_{ADC}$ is the *Figure of Merit* of the ADC and determines how much energy consumption increases per *enob*. For ADCs suitable to the applications under consideration, $FOM$ is typically around 10–50 fJ/conv-step [5].

Energy consumption at the output of the amplifier is given by

$$E_{gm,out} = \frac{V_{DD}I_D}{2f_{sig}}, \tag{3.10}$$

where $V_{DD}$ is the supply voltage and the output current $I_D$ ranges between $I_{D,min}$ and $9I_{D,min}$, assuming a 2–3 stage amplification. $I_{D,min}$ can be estimated using a $gm/I_D$ based design approach: $I_{D,min} = 2\frac{gm_{inp,min}}{\eta}$, where $\eta$ depends on the biasing of the input transistors. The value of $gm_{inp,min}$ will be defined to meet the noise tolerance specifications of the front-end. A common design guideline dictates that the noise of the amplifier should be smaller than the ADC's quantization noise, such that the ADC can yield a meaningful output:[2]

$$V_{n,gm,out}^2 < V_{n,quant,ADC}^2.$$ (3.11)

The equation above can be rewritten as

$$\left(1 + \frac{C_i}{C_f}\right)\frac{2\alpha\gamma kT}{C_L} < \frac{1}{12}\frac{V_{ref}^2}{2^{2enob}},$$ (3.12)

which allows to set a minimum limit on the load-capacitor $C_L$, as well as to determine the minimum transconductance needed to reach the desired *enob*:

$$C_{L,min} = \left(1 + \frac{C_i}{C_f}\right)\frac{24\alpha\gamma kT}{V_{ref}^2}2^{2enob}$$ (3.13)

$$gm_{inp,min} = \left(1 + \frac{C_i}{C_f}\right)2\pi f_{sig}C_{L,min}.$$ (3.14)

Therefore, under the condition in Eq. (3.11), the total front-end energy consumption as a function of *enob* can be written as

$$E_{tot}^{1stage} = 48\pi\frac{\alpha\gamma}{\eta}kT\left(1 + \frac{C_i}{C_f}\right)^2\frac{V_{DD}2^{2enob}}{V_{ref}^2} + 2^{enob}FOM_{ADC}.$$ (3.15)

For a quality scalable front-end $S$, the sensing cost is a function of *enob* and is based on Eq. (3.15):

$$C_S(S) = 2^{enob_S} \cdot (\beta_1 \cdot 2^{enob_S} + \beta_2),$$ (3.16)

where $enob_S$ is the SNR of sensor front-end $S$, and $\beta_1$ and $\beta_2$ are factors based on Eq. (3.15) that remain constant and are given by: $\beta_1 = 48\pi\frac{\alpha\gamma}{\eta}kT\left(1 + \frac{C_i}{C_f}\right)^2\frac{V_{DD}}{V_{ref}^2}$ and $\beta_2 = FOM_{ADC}$. The values for each variable within these two terms were discussed above and follow common design practices and standards [4].

Note that the models derived throughout this section represent the pure physical (noise limited) properties of the circuits at the sensor front-end, independent of

---

[2]Note that the methodology proposed in Chap. 4 aims to exploit the energy consumption savings from increasing $V_{n,gm,out}^2$. Therefore, it is assumed in that chapter that the amplifier's noise dominates over the quantization noise.

the technology node[3] used. In practice, the technology process sets a limit to the minimum achievable capacitance, which is denoted by $C_{min}$. Equation (3.15) does not follow the physical limits when the desired SNR is so low that it dictates that $C_L < C_{min}$ (in that situation $C_L$ in Eq. (3.3) would have to be equal to $C_{min}$). For example, the energy consumption of circuits with a technology of 90 nm ceases to decrease exponentially for SNR lower than 9 *enob* [6]. For smaller technology nodes, this SNR threshold, setting the noise-dependent energy consumption scaling, is also lower.

## 3.3 Feature Extraction Costs

As mentioned in Sect. 2.4.1, most embedded machine learning implementations extract features in the digital domain, once the incoming sensory signal has been processed and digitized by the mixed-signal front-end. For this and other digital domain functionality, the applications considered in this book rely on lightweight processors, mainly embedded CPUs to implement the required digital blocks.

At a high level, a typical embedded hardware architecture entails two components: an on-chip main memory, which commonly houses the algorithm's parameters; and a processing unit, where operations are performed and intermediate values are cached in a local memory. Resource consumption of digital feature extraction is primarily determined by (1) how many features are extracted and how complex they are in terms of the type (additions or multiplications) and number of required operations and (2) what precision is used to represent them and perform the necessary arithmetic operations.

**Relative Energy Consumption per Operation Type** For most processors, one can identify relative trends in the energy consumption of the different available operations. This energy consumption scaling is determined by the particular characteristics of the hardware implementation in terms of architecture and semiconductor technology node. For example, the energy cost estimations from the benchmarks in [7] indicate that, for an embedded CPU in 45 nm CMOS technology, floating point multipliers can be three to four times more expensive than adders and that caching and memory transactions can be tens and even hundreds of times more expensive.

**Low Precision and Energy Consumption** Reduced precision also has an impact on energy consumption of arithmetic operators and memory transactions. Energy scales, due to, for example, reduced critical paths in the operators [8], or the need for less digital gates [9]. Table 3.1 shows post-synthesis energy models for fixed point and floating point adders and multipliers in 65 nm CMOS in terms of number of

---

[3]This is an industry standard that designates the process' transistor gate size, and is typically given in nanometers. Today's devices use nodes ranging from 65 to 5 nm.

**Table 3.1** Energy models for arithmetic operators at 1 V for 65 nm CMOS [10]

| Operator | Energy (fJ) |
|----------|-------------|
| Fixed-point adder | $7.8 \cdot nb$ |
| Fixed-point multiplier | $1.9 \cdot nb^2$ |
| Float-point adder | $44.74 \cdot (nb + 1)$ |
| Float-point multiplier | $2.9 \cdot (nb + 1)^2 \cdot \log(nb + 1)$ |

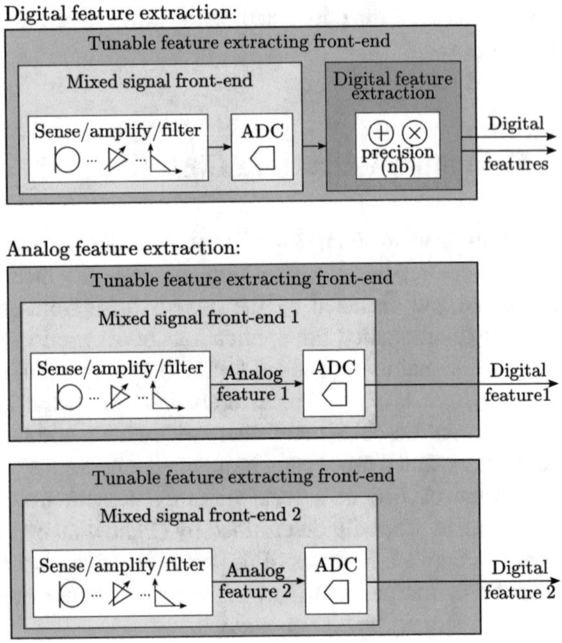

**Fig. 3.3** Feature extraction in the digital domain and the analog domain

bits $nb$.[4] The coefficients in these models were obtained by fitting functions to post-synthesis simulations of the different operators. In general, energy consumption of multipliers scales quadratically with the number of bits and linearly for adders.

The cost of features extracted in the digital domain is therefore determined by the type of operation and the precision scaling described above. This will be formalized in Sect. 3.3.1. Note, however, that there are systems that extract features in the analog domain, before the digitization by the ADC. The guiding principle of this approach is early information discrimination, which can be achieved by extracting features as close as possible to the sensor. This reduces the rate at which digital data is produced and can eliminate the need to extract features in the digital domain. The feature extraction cost of this approach is described in Sect. 3.3.2 and follows the same energy versus quality scaling derived for the mixed-signal front-end. The two feature extraction approaches are illustrated by Fig. 3.3.

---

[4]These energy models were included in [10] and proposed by Nimish Shah.

### 3.3.1 Digital Feature Precision Scaling

For the definition of $C_F$, digital feature extraction is assumed to take place with a series of multiply–accumulate (MAC) operations. For example, the calculation of mean or variance requires one MAC operation, while a 64 point FFT requires six. Since the energy consumption of multipliers is dominant (they scale quadratically with the number of bits as shown in Table 3.1), the cost of extracting features is

$$C_F(F) = \beta \cdot enob_F^2, \tag{3.17}$$

where $\beta$ is a feature-specific cost factor, based on the number of required MAC operations, and $enob$ is the effective number of bits procured by the ADC, with which MAC operations can be performed.[5]

### 3.3.2 Analog Feature Precision Scaling

Some analog sensor front-ends are designed to extract features [1], in addition to the amplification discussed in Sect. 3.2.1. This follows the same exponential energy consumption scaling. Thus, for this type of systems, the sensor interfacing costs are given by

$$C_{F,\text{analog}}(F) = \beta \cdot 2^{enob_F}, \tag{3.18}$$

where $\beta$ are feature-specific energy costs. Section 4.6, for example, will consider a system that extracts features in the analog domain, and the feature-specific factor $\beta$ scales exponentially with each of the 16 features available. Note that this cost assumes that a feature is extracted from each sensory chain, as shown in Fig. 3.3.

## 3.4 Inference Costs

In addition to the feature extraction block, the classifier and control blocks of Fig. 3.1 are implemented in the digital domain.

In particular, the machine learning task is assumed to be implemented in an embedded CPU entailing two main components: an on-chip main memory housing the model's parameters; and a processing unit where arithmetic operations are

---

[5]Equation (3.17) assumes that the $nb$ used in the digital domain are equal to the $enob$ provided by the ADC. This equation represents a lower bound: in practice $nb$ tends to be a few bits higher than $enob$, which can be taken into consideration within the formulation of $\beta$.

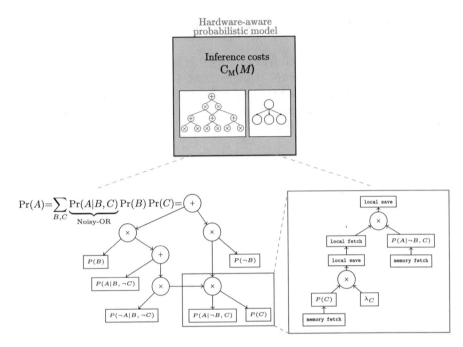

**Fig. 3.4** Inference cost, evaluated on a Probabilistic Circuit. Note that Bayesian networks can be compiled to Probabilistic Circuit representations

performed and intermediate values are cached in a local memory. Inference cost on model $\mathcal{M}$ is defined as

$$C_M(\mathcal{M}, nb) = C_+(nb) + C_\times(nb) + C_{mem}(nb) + C_{cache}(nb), \qquad (3.19)$$

where the cost of each type of operation (addition, multiplication, memory exchange, and cache exchange) is a function of the number of bits $nb$ of precision.

When the machine learning task is executed on a Probabilistic Circuit (PC), inference costs are straight forward to compute: recall that PCs (see Sect. 2.3) are computational graph representations comprising multiplications, additions, and parameters, which can be readily used to compute, for example, marginal probabilities. Performing an upward pass on an PC involves the following actions (see Fig. 3.4): (1) fetching parameters from the main memory, (2) performing arithmetic operations, consisting of additions and multiplications, (3) caching intermediate values in a local memory (e.g. register file or low-level cache), and (4) fetching intermediate values from local memory, as needed. Note that this book assumes that the local cache size is sufficiently large to store intermediate values, but not large enough to store parameters. However, for some learned circuits, there are about as many parameters as edges, so, depending on the local memory size, one might need to store intermediate values also in main memory. Inference cost is therefore defined from the structure of Probabilistic Circuit $\kappa$:

**Table 3.2** Inference costs

| Operation | At 64 bits | Operation cost |
|---|---|---|
| $C_{mem}$ | 1 | $\phi_{mem} = \gamma_{mem} \cdot nb$ |
| $C_{cache}$ | 0.2 | $\phi_{cache} = \gamma_{cache} \cdot nb$ |
| $C_\times$ | 0.6 | $\phi_\times = \gamma_\times^2 \cdot nb^2 \cdot \log(nb)$ |
| $C_+$ | 0.1 | $\phi_+ = \gamma_+ \cdot nb$ |

$$C_+(nb) = \sum_{k \in \kappa}[k = +] \cdot \phi_+(nb),$$

$$C_{mem}(nb) = \sum_{k \in \kappa}[k \neq + \wedge k \neq \times] \cdot \phi_{mem}(nb),$$

$$C_\times(nb) = \sum_{k \in \kappa}[k = \times] \cdot \phi_\times(nb),$$

$$C_{cache}(nb) = \sum_{k \in \kappa}[k = + \vee k = \times] \cdot \phi_{cache}(nb) \cdot (|children(k)| + 1),$$

where $k$ denotes a node in $\kappa$, $[\beta]$ is equal to 1 when $\beta$ is true, and the equality in $\beta$ holds when node $k$ matches the operation type (i.e. $+$ or $\times$). The function $\phi(\cdot)$ describes the effective cost of the particular operation and can be derived from empirical benchmarks, customized to the target hardware [7, 10]. Finally, all these costs scale with the precision in number of bits used to represent parameters and perform arithmetic operations ($nb$), which is typically the same for all nodes in the Probabilistic Circuit. Recall from the discussion above that arithmetic operations are assumed to store their intermediate values in local memory, and that they fetch the values they require from local memory as well. Therefore, the cost of local memory $C_{cache}$ considers the number of values that must be fetched ($|children(k)|$) to perform the arithmetic operation in question, as well as the result to save ($+1$). Main memory cost $C_{mem}(nb)$ is assumed to only be incurred by leaf nodes, as they represent univariate distributions parametrized by $\theta_L$. Note also that, in practice, the parameters corresponding to sum nodes $\theta_S$ (see Sect. 2.3) are also implemented as leaves in the PC and fetching them is therefore considered in the formulation of $C_{mem}(nb)$.

This book considers the operation-cost definitions of Table 3.2, based on the relative energy consumption provided by the benchmarks in [7, 10]. The definition of $\gamma$ in these costs is based on the energy models from Table 3.1 and is provided at a baseline precision of 64 floating point bits, since it is the standard IEEE representation in software environments.

## 3.5 Dynamic Tuning Costs for Run-Time Strategies

The control block implementing the policy for run-time tuning shown in Fig. 3.1 is also implemented in digital hardware, and its hardware-aware cost $C_T$ is therefore also a function of the number and type of elementary operations, and the number of bits. Chapter 6 proposes strategies that extract the Pareto-optimal trade-off at run-time, relying on the models proposed in Chaps. 4 and 5. As will be discussed in Chap. 6, timing aspects must also be considered when estimating the energy consumption of run-time implementations. In general, the control block applies the policy at a lower rate than features are extracted and inference must be made, so energy in this block is also consumed at a lower rate.

## 3.6  Types of Systems Considered in This Book

Depending on the type of system, energy consumption tends to be dominant in particular blocks of the sensory embedded pipeline. For example, for always-on sensor systems, energy consumption at the front-end is dominant over the digital inference blocks' energy consumption. Other types of IoT nodes suppose more complex digital processing, and therefore inference energy consumption is dominant.

Table 3.3 provides an overview of the types of systems considered throughout the following chapters, and their definition of hardware-aware cost. Chapter 4 focuses on systems with dominant sensor front-end blocks, where the quality of the processed signal and extracted features can be tuned and traded-off for energy consumption [1, 11]. As shown in Table 3.3, three use cases of scalable front-ends are considered: tunable mixed-signal systems, systems with tunable digital feature extraction, and systems that extract features in the analog domain. Chapter 5 considers devices with general purpose, non-scalable sensor front-ends, and more advanced digital blocks implementing the machine learning task. Finally, Chap. 6 considers the implications of run-time tuning strategies implemented in the digital domain. Two use cases are considered here: dynamic tuning of scalable mixed-signal front-ends, and dynamic scaling of general purpose systems, implementing the machine learning task on a PC.

**Table 3.3**  Hardware-aware cost used throughout this book

| Chapter | System type | $C_{HA}$ definition |
|---|---|---|
| Chapter 4 | Tunable mixed-signal front-end | $C_S$ |
| | Digital quality scaling at the front-end | $C_F$ |
| | Analog feature extraction | $C_{F,analog}$ |
| Chapter 5 | General purpose, advanced digital processing | $C_S + C_F + C_M$ |
| Chapter 6 | Tunable mixed-signal front-end at run-time | $C_S + C_M + C_T$ |
| | General purpose, run-time | $C_S + C_F + C_M + C_T$ |

## 3.7 Conclusion

This chapter introduced the notion of hardware-aware cost, an abstraction of system-wide energy consumption, which is used for the optimization strategies proposed throughout this book. It is shown how this metric can be customized to represent scalable aspects of different types of systems, providing the definitions that each subsequent chapter will rely on.

## References

1. K.M. Badami, S. Lauwereins, W. Meert, M. Verhelst, A 90 nm CMOS, power-proportional acoustic sensing frontend for voice activity detection. IEEE J. Solid-State Circ. **51**(1), 291–302 (2016)
2. W. Dehaene, R. Uytterhoeven, C. Nieto Taladriz Moreno, B. Vanhoof, Dealing with the energy versus performance tradeoff in future CMOS digital circuit design, in *NANO-CHIPS 2030* (Springer, New York, 2020), pp. 89–115
3. L. Galindez, K. Badami, J. Vlasselaer, W. Meert, M. Verhelst, Dynamic sensor-frontend tuning for resource efficient embedded classification. IEEE J. Emerg. Select. Top. Circ. Syst. **8**(4), 858–872 (2018)
4. M.J. Pelgrom, Analog-to-digital conversion, in *Analog-to-Digital Conversion* (Springer, New York, 2013), pp. 325–418
5. B. Murmann, ADC performance survey 1997–2017 (2017). http://web.stanford.edu/~murmann/adcsurvey.html
6. T. Sundstrom, B. Murmann, C. Svensson, Power dissipation bounds for high-speed Nyquist analog-to-digital converters. IEEE Trans. Circ. Syst. I: Regul. Pap. **56**(3), 509–518 (2009)
7. M. Horowitz, 1.1 Computing's energy problem (and what we can do about it), in *2014 IEEE International Solid-State Circuits Conference Digest of Technical Papers (ISSCC)*, February 2014, pp. 10–14
8. B. Moons, M. Verhelst, DVAS: dynamic voltage accuracy scaling for increased energy-efficiency in approximate computing, in *2015 IEEE/ACM International Symposium on Low Power Electronics and Design (ISLPED)* (IEEE, New York, 2015), pp. 237–242
9. J.M. Rabaey, A.P. Chandrakasan, B. Nikolić, *Digital Integrated Circuits: A Design Perspective*, vol. 7 (Pearson Education, Upper Saddle River, NJ, 2003)
10. N. Shah, L.I. Galindez Olascoaga, W. Meert, M. Verhelst, ProbLP: a framework for low-precision probabilistic inference, in *Proceedings of the 56th Annual Design Automation Conference 2019*, 2019, pp. 1–6
11. J. De Roose, H. Xin, M. Andraud, P.J. Harpe, M. Verhelst, Flexible and self-adaptive sense-and-compress for sub-microwatt always-on sensory recording, in *ESSCIRC 2018-IEEE 44th European Solid State Circuits Conference (ESSCIRC)* (IEEE, New York, 2018), pp. 282–285

# Chapter 4
# Hardware-Aware Bayesian Networks for Sensor Front-End Quality Scaling

As discussed in Chaps. 2 and 3, hardware design choices in the sensor front-end are driven by target output quality and precision. In turn, these design choices determine the front-end's energy consumption. Whereas most devices are fabricated to meet fixed quality specifications, some sensor front-ends comprise adjustable components that allow them to tune their target quality, and, as a consequence, their energy consumption.

This chapter leverages front-end tunability to systematically trade-off limited accuracy losses for cost savings. The proposed strategy relies on a hardware-aware model representing the implications of different qualities and precisions of the sensor front-end. Thus, a single model can operate at a wide range of cost versus accuracy performance. For example, the model can be used to compare the accuracy and cost of front-ends operating at 10 *enob* versus 8 *enob* SNR. As such, it can aid in the selection of an appropriate front-end configuration to deploy at application time.

The proposed model in question is the ns-BN. This model relies on the graphical and probabilistic aspects of Bayesian networks to represent features extracted with different levels of quality and precision, as procured by the scalable hardware. The premise behind the ns-BN is that one can express the relation among different levels of signal quality by means of a probability distribution, the choice of which is guided by statistical models from circuit design and signal processing theory (some were described in Chap. 3). Encoding these distributions with a Bayesian network enables probabilistic inference over the available signal quality variations. A single model can thus aid in evaluating the impact that different amounts of hardware-induced noise may have on the classification task. This bypasses the need to produce an array of models trained with noisy data in trying to establish the effect of reduced sensory quality on task performance.

This chapter is organized as follows. Section 4.1 formalizes the ns-BN model and its use in classification tasks. Then, Sect. 4.2 introduces a technique that leverages the multi-quality evaluation aspects of the ns-BN to search the cost versus accuracy

© The Author(s), under exclusive license to Springer Nature Switzerland AG 2021
L. I. Galindez Olascoaga et al., *Hardware-Aware Probabilistic Machine Learning Models*,
https://doi.org/10.1007/978-3-030-74042-9_4

trade-off space and extracts the Pareto-optimal front within it. This chapter considers three use cases of the ns-BN, each targeting different scalable aspects of sensor front-ends, as explained in Sect. 4.3. The first use case, corresponding to tunable mixed-signal front-ends, is introduced in Sect. 4.4 and is based on the following publication:

> Galindez, L., Badami, K., Vlasselaer, J., Meert, W., and Verhelst, M. (2018). Dynamic Sensor-Frontend Tuning for Resource Efficient Embedded Classification. *IEEE Journal on Emerging and Selected Topics in Circuits and Systems*, 8(4), 858–872.

Section 4.5 introduces the second use case, which focuses on tunable digital feature extraction and is based on:

> Galindez Olascoaga, L. I., Meert, W., Bruyninckx, H., and Verhelst, M. (2016). Extending Naive Bayes with Precision-Tunable Feature Variables for Resource-Efficient Sensor Fusion. *In 2nd AI-IoT Workshop collocated with ECAI 2016* (Vol. 1724, pp. 23–30). CEUR-WS.

The third use case, introduced in Sect. 4.6, focuses on systems that can extract features in the analog domain and is based on:

> Galindez Olascoaga, L. I., Badami, K., Pamula, V. R., Lauwereins, S., Meert, W., and Verhelst, M. (2016). Exploiting System Configurability Towards Dynamic Accuracy-Power Trade-Offs in Sensor Front-Ends. *In 50th Asilomar Conference on Signals, Systems and Computers* (pp. 1027–1031). IEEE.

Finally, related work is listed in Sect. 4.7 and a concluding discussion is provided in Sect. 4.8.

## 4.1   Noise-Scalable Bayesian Network Classifier

This section formally defines the properties of the ns-BN and specifies how to use it for inference and classification.

### 4.1.1   Model

The ns-BN consists of a Bayesian network classifier that has been extended with nodes representing various (circuit-)noise prone and low-precision versions of each feature. Figure 4.1a shows an example of a five-feature ns-BN, where $F_i$ denotes the *noiseless* feature $i$ and $F_i'$ denotes its *noisy* version.[1] The proposed structure encodes the following joint probability distribution over the class variable $C$ and

---

[1] Features are described as *noiseless* when they have been extracted by a signal processing system that meets the highest-quality or baseline-quality specifications of the application of interest. *Noisy* features are prone to circuit noise or extracted at a low precision.

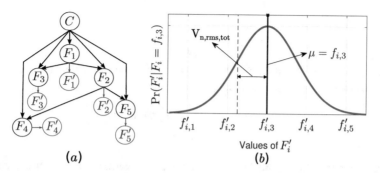

**Fig. 4.1** (**a**) Example of the noise-scalable Bayesian network classifier (ns-BN) created by extending a Tree Augmented Naive Bayes classifier with noisy feature versions. (**b**) An example of the probabilistic relation between feature $F_i$ and its noisy version $F_i'$, given by a Gaussian distribution. Taken from [1]

the noiseless and noisy feature sets $\mathbf{F} = \{F_1, \ldots, F_n\}$ and $\mathbf{F}' = \{F_1', \ldots, F_n'\}$, respectively:

$$\Pr(C, \mathbf{F}, \mathbf{F}') = \prod_{i=1}^{n} \Pr(F_i'|F_i) \cdot \Pr(F_1, \ldots, F_n, C) , \qquad (4.1)$$

where the distribution $\Pr(F_1, \ldots, F_n, C)$ encodes a Bayesian network classifier over $\mathbf{F}$ and $C$. This could, for example, be a TAN classifier, as shown in Fig. 4.1a. The distributions $\Pr(F_i'|F_i)$ are selected to represent the noise properties of the targeted tunable hardware. Figure 4.1b encodes circuit noise subjected on $F_i'$ through a Gaussian[2] distribution with mean $\mu = F_i$ and standard deviation $\sigma$ proportional to the sensor front-end's noise (and therefore, inversely proportional to its SNR).

Like in the example above, the parameters that define the distribution $\Pr(F_i'|F_i)$ are proportional to the amount of noise tolerated, which, as mentioned before, is tunable in the type of systems targeted in this chapter. Each feature $F_i$, or each group of features extracted from a common sensory chain $\mathbf{S_i}$ (see Fig. 4.3) can be subjected to $h$ user defined levels of noise. Therefore, each distribution $\Pr(F_i'|F_i)$ can be parametrized by $h$ different parameters sets $\Theta_i = \{\Theta_i^{(1)}, \ldots, \Theta_i^{(h)}\}$. What this means for the graphical component of the ns-BN is that the same node $F_i'$ can be (re)used with different hardware noise specifications.

---

[2]Note that the Gaussian distribution is evaluated over the values of $F'$ and stored in a table. At inference time, the corresponding discrete value of the Gaussian is fetched from this table.

### 4.1.2   Inference

The ns-BN can be used to answer queries of the form $\Pr(C|\mathbf{f})$, where $\mathbf{f}$ is an instantiation of the features. Classification is performed by selecting the class $C$ that maximizes $\Pr(C|\mathbf{f})$ as follows:

$$\Pr(C|\mathbf{f}) \sim \sum_{\mathbf{F}\backslash\mathbf{f}} \sum_{\mathbf{F}'\backslash\mathbf{f}} \Pr(C, \mathbf{F}, \mathbf{F}'), \qquad (4.2)$$

where the summations indicate that unobserved features are marginalized. As explained at the beginning of this chapter, the ultimate goal of the ns-BN model is to aid in searching the accuracy versus cost trade-off space by enabling inference on feature observations with varying levels of hardware-induced quality. Thus, evidence can be available over both noisy features and noiseless features during this off-line search. In the use cases to follow, noiseless features $\mathbf{F}$ are often assumed to remain unobserved, in which case they are always marginalized.

As such, the ns-BN can be used to evaluate classification accuracy under different levels of hardware-induced noise (including the situation where a sensor is turned off and features from that sensor are not extracted). The concepts described in Chap. 3 allow to evaluate the resource consumption and define the hardware-aware cost under these setting variations. This is possible because each $F'$ node can be used to represent several gradations of noise, as described in the previous section.

Sections 4.4, 4.5, and 4.6 propose three use cases of the ns-BN, which address different sources of tunable hardware-induced noise. These sections show how the ns-BN can be used to explore the cost versus accuracy trade-off space available for the use case or application of interest. Prior to this, Sect. 4.2 proposes a strategy that extracts the Pareto-optimal front within the space explored by the ns-BN. The goal of this strategy is to define the sensor front-end configurations that, for a given cost, will attain the highest accuracy possible, or vice versa. The developer or the user can then define a desired cost and accuracy and select from these Pareto-optimal configurations accordingly, such that the system can meet the desired performance at run-time.

## 4.2   Local Pareto-Optimal Feature Quality Tuning

As discussed at the beginning of this chapter, one of the roles of the ns-BN is aiding in the systematic search of the cost versus accuracy trade-off space. The final goal is to identify sets of feature configurations that are Pareto-optimal within this explored space. The ns-BN's structure allows to represent each feature—or sets of features per sensory chain—at multiple hardware-induced levels of quality or noise tolerance. Furthermore, the model is tailored to the targeted hardware by parametrizing the distribution $\Pr(F_i'|F_i)$ in accordance to the properties of the

---

**Algorithm 1:** SCALEFEATURENOISE(BN, $acc_{target}$, $cost_{target}$, OF, $\theta_{init}$)

---

**Input:** BN: trained model, $acc_{target}$, $cost_{target}$: target accuracy and cost, OF: user defined objective function, $\theta_{init}$: initial feature-wise configuration

**Output:** $\langle \mathcal{T}, \textbf{acc}, \textbf{cost} \rangle$: selected feature configurations, their accuracy and cost

1   $\theta_{select} \leftarrow \theta_{init}$

2   $acc_{select} \leftarrow \text{Acc}(\text{BN}, \theta_{select}, \mathcal{F}_{valid})$

3   $cost_{select} \leftarrow \text{CostF}(\theta_{select})$

4   $\mathcal{T} \leftarrow \theta_{select}, \langle \textbf{acc}, \textbf{cost} \rangle \leftarrow \langle acc_{select}, cost_{select} \rangle$

5   **while** $acc_{select} \leq acc_{target} \wedge cost_{select} \geq cost_{target}$ **do**

6     **foreach** $\theta_i^{(v_i)} \in \theta_{select}$ **do**

7       **if** $\theta_i^{(v_i)} == \max(\Theta_i)$ **then**

8         $\theta_{cand,i} \leftarrow \theta_{select}$

9       **else**

10         $\theta_{cand,i} \leftarrow \theta_{select} \setminus \theta_i^{(v_i)} \wedge \theta_i^{(v_i+1)}$

11         $acc_{cand,i} \leftarrow \text{Acc}(\text{BN}, \theta_{cand,i}, \mathcal{F}_{valid})$

12         $cost_{cand,i} \leftarrow \text{CostF}(\theta_{cand,i})$

13     $\theta_{select} \leftarrow \underset{\theta \in \theta_{cand}}{\text{argminOF}}(acc_{cand}, cost_{cand})$

14     $acc_{select} \leftarrow \text{Acc}(\text{BN}, \theta_{select}, \mathcal{F}_{valid})$

15     $cost_{select} \leftarrow \text{CostF}(\theta_{select})$

16     $\mathcal{T}.\text{insert}(\theta_{select}), \textbf{acc}.\text{insert}(acc_{select}), \textbf{cost}.\text{insert}(cost_{select})$

17     **if** $\forall i : \theta_i^{(v_i)} == \max(\Theta_i)$ **then**

18       **break**

19 **return** $\langle \mathcal{T}, \textbf{acc}, \textbf{cost} \rangle$

---

tunable component's noise: for example, by a Gaussian distribution for mixed-signal noise (Sect. 4.4) and by a deterministic distribution for quantization noise (Sect. 4.5). In addition, the model's structure encodes independence assumptions among noisy features, which reflects the fact that many devices include independent processing chains per sensor.[3] Searching the full trade-off space and finding the globally optimal configuration becomes a combinatorial problem with exponential time complexity. This motivates the proposal of a local search strategy and the subsequent extraction of the Pareto front within this locally searched space.

The proposed strategy, SCALEFEATURENOISE (Algorithm 1), is based on a greedy neighborhood search that iteratively attempts to reduce the quality of each feature—or sensory chain-wise feature set—until the target performance is met. The target performance is defined by the developer or the user and given in terms of accuracy or hardware-cost ($acc_{target}$, $cost_{target}$), without loss of generality.[4] In addition to target performance, the user can determine the objective function (OF)

---

[3] The assumption made here is that the noise arising at each of them is uncorrelated.

[4] Performance metrics other than accuracy and cost models that consider resources other than energy consumption can also benefit from the ns-BN and the search strategy.

---

**Algorithm 2:** GETPARETO($\mathcal{T}$,acc,cost)

---

**Input**: $\mathcal{T}$, **acc**, **cost**: Configuration set, their accuracy and cost.
**Output**: $\mathcal{T}^*$, **acc**$^*$, **cost**$^*$: Pareto optimal configurations, their accuracy and cost.
1  $\langle$**cost**$^*$, **acc**$^*$, $\mathcal{T}^*$$\rangle$ ← $\langle$\{\}, \{\}, \{\}$\rangle$ ;
   /* Sort according to ascending cost                      */
2  $\langle$**cost**, **acc**, $\mathcal{T}$$\rangle$ ← $sorted(\langle$**cost**, **acc**, $\mathcal{T}$$\rangle)$;
3  $j \leftarrow |\mathcal{T}| + 1$;
4  **while** $j > 0$ **do**
5      $j \leftarrow \arg\max \mathbf{acc}_{0:j}$
6      $\mathcal{T}^*$.insert($\boldsymbol{\theta}_j$)
7      **acc**$^*$.insert($acc_j$)
8      **cost**$^*$.insert($cost_j$)
9      $j \leftarrow j - 1$
10  **return** $\mathcal{T}^*$, **acc**$^*$, **cost**$^*$

---

to optimize at each iteration, which defines how accuracy and cost will be traded off during the search.

The algorithm is initialized to represent the highest quality feature-wise setting $\boldsymbol{\theta}_{init}$, although it can also be initialized to other quality conditions of interest, as will be discussed in Chap. 6, which proposes a run-time noise tuning strategy. This noise configuration is used for the initial estimation of accuracy and cost ($acc_{select}$ and $cost_{select}$ in lines 2 and 3). Note that the strategy assumes that a validation dataset $\mathcal{F}_{valid}$ is available for accuracy evaluation. At each iteration, the algorithm tries to reduce the quality of feature $F_i$: the current noise tolerance value $v_i$ is increased by one unit (line 10), unless the previously selected configuration is already at its highest possible noise tolerance (line 8).[5] This loop (line 6), attempting to reduce each feature's quality, produces as many quality reduction candidates as there are features. The algorithm then proceeds to select the candidate that optimizes the objective function OF, defined in terms of the accuracy and cost ($acc_{cand}$, $cost_{cand}$). The experiments throughout Sects. 4.4, 4.5, and 4.6 consider different objective functions for each use case.

Algorithm 2 outputs the set of locally optimal feature-noise configurations $\mathcal{T}$, their accuracy and cost. The local Pareto-optimal front within this set can then be extracted with any convex hull algorithm, such as the one described in Algorithm 2, which iteratively looks for the configuration $\boldsymbol{\theta}_j$ that results in the maximum accuracy at the currently considered $cost_j$. Thus, the algorithm outputs the noise configuration $\mathcal{T}^*$ corresponding to each local Pareto-optimal operating point (**acc**$^*$, **cost**$^*$).

Section 4.3 introduces three use cases of the ns-BN and shows how the SCALEFEATURENOISE strategy leverages the properties of this model to extract the front-end quality configurations that lead to local Pareto-optimal performance.

---

[5]For certain use cases, the highest possible noise tolerance is equivalent to pruning the feature.

## 4.3  Use Cases of the ns-BN: Introduction

One of the qualities of the ns-BN is that it allows representing the effects of a variety of hardware noise sources on the extracted features. This chapter considers three use cases, each targeting a different type of sensor front-end as shown in Fig. 4.2, which also highlights in red the sub-systems that enable noise scalability.

The first use case, shown at the top of Fig. 4.2, considers sensor front-ends with tunable amplification circuitry and, as a consequence, scalable mixed-signal noise (defined as $V_{n,rms,tot}^2$ in Eq. (3.5)). This use case relies on the equations derived in Sect. 3.2.1 to define the ns-BN and associated hardware-costs and will be discussed in detail in Sect. 4.4.

The second use case, discussed in Sect. 4.5, exploits the scalability of the ADC and the digital feature extraction block, as shown in the middle of Fig. 4.2. This use case defines the ns-BN in terms of quantization noise and uses the energy models in Sect. 3.3 to determine the costs.

The third use case is discussed in Sect. 4.6 and is inspired by the ultra-low power voice activity detection (Voice Activity Detection (VAD)) system proposed in [2], which extracts features in the analog domain. The ns-BN and associated costs are defined accordingly.

Each of the use cases above was evaluated empirically on a variety of publicly available datasets. The experiments target the common goal of identifying the quality configurations that lead to local Pareto-optimal accuracy versus cost.

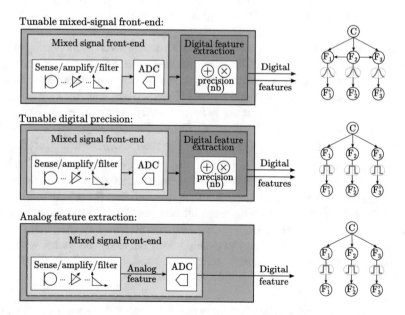

**Fig. 4.2** Left: scalable blocks targeted by the use cases. Right: Structure and parametrization of the ns-BN used for experiments

Moreover, they showcase the versatility of the ns-BN as well as the fact that the SCALEFEATURENOISE strategy is capable of considering different sub-systems within the front-end, from the purely analog, through the mixed signal, to the digital blocks.

## 4.4   First Use Case: Mixed-Signal Quality Scaling

The first use case of the ns-BN exploits the quality versus energy consumption trade-off that arises at the mixed-signal front-end highlighted in red in Fig. 4.3a. As discussed in Sect. 3.2.1, the SNR describes the front-end's performance in terms of its tolerance to random[6] circuit noise. One can thus model the relation between a *noiseless* signal processing system and a *noisy* one by means of a Gaussian distribution, with standard deviation inversely proportional to SNR.

This use case considers applications that extract features in the digital domain, once the sensory signal has been processed by the front-end, as shown in Fig. 4.3a. It is assumed that the relation between *noisy* and *noiseless* features can be described by a Gaussian distribution in the same way circuit noise is, as shown in Fig. 4.3b. This assumption is based on the fact that most operations taking place at the digital feature extracting block consist of additions and multiplications (as explained in Chap. 3). Additions and multiplications between two signals with Gaussian noise result in a signal with Gaussian noise, i.e. $a + noise_a + b + noise_b = c + noise_c$. Therefore, in the ns-BN corresponding to this use case, the distribution $\Pr(F_i'|F_i)$ is Gaussian with mean $\mu = F_i$ and standard deviation equal to $V_{n,rms,tot}$, which is the total noise of the front-end and is inversely proportional to SNR.

As explained in Sect. 4.1, the ns-BN allows $h$ possible levels of increasing quality degradation per feature or set of features extracted from a common sensory chain. For instance, in the system shown in Fig. 4.3a there are two sensor-wise feature sets, defined as $\mathbf{S}_1 = \{F_1, F_2\}$ and $\mathbf{S}_2 = \{F_3, F_4\}$. Here, features $F_1$ and $F_2$ are subjected to the same level of circuit noise since they are extracted by a common front-end. The same is true for features $F_3$ and $F_4$. The parameter set of the ns-BN in this example is thus given by $\Theta = \{\Theta_{S1}, \Theta_{S2}\}$, with $\Theta_{S1} = \{\theta_{S1}^{(1)}, \ldots, \theta_{S1}^{(h)}\} = \{enob_{S1}^{(1)}, \ldots, enob_{S1}^{(h)}\}$ and $\Theta_2 = \{\theta_{S2}^{(1)}, \ldots, \theta_{S2}^{(h)}\} = \{enob_{S2}^{(1)}, \ldots, enob_{S2}^{(h)}\}$. Note that, for this example, two candidates (one per sensory chain) are considered by the for loop in line 6 of Algorithm 1, even though there are four features in total.

The ns-BN allows to evaluate the trade-off between classification accuracy under varying circuit noise tolerance and energy-cost. Recall from Eq. (3.8) and the cost definitions in Chap. 3 that energy consumption of the front-end scales with SNR (given in terms of *enob*). The search strategy discussed in Sect. 4.2 can therefore be

---

[6]The thermal noise arising at the amplifier can be assumed to be random, while quantization noise from the ADC is assumed to be uniform in most applications, but can be assumed to be random at very high resolutions (>10 bits) [3].

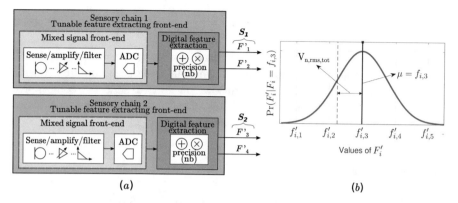

**Fig. 4.3** (**a**) Block diagram of the tunable mixed-signal sensor front-end. (**b**) Gaussian distribution parametrizing the relation between *noisy* and *noiseless* features

used to extract the Pareto-optimal trade-off front in this accuracy versus energy-cost trade-off space.

### 4.4.1   Experiments for Mixed-Signal Quality Scaling

The experiments in this section target systems with noise-scalable sensor front-ends, where the energy consumption of the mixed-signal sub-block dominates (see Sect. 3.2.1). The goal of the experiments herewith is therefore to determine the level of SNR per sensory chain that will consume minimal energy-cost for a certain target accuracy or vice versa. In other words, determine the sensory-chain SNR configurations that map to the local cost versus accuracy Pareto front.

#### Cost Function and Objective Function Definition

As discussed in Sect. 3.6 and shown in Table 3.3, the sensing costs $C_S$ are dominant for this use case. Therefore, for the experiments herewith, the definition of the cost function in line 6 of Algorithm 1 is based on Eq. (3.16) and given by

$$\text{CostF}(\boldsymbol{\theta}) = \sum_{S \in \mathbf{S}} C_S(S) = \sum_{S \in \mathbf{S}} 2^{\theta_S} \cdot (\beta_1 \cdot 2^{\theta_S} + \beta_2), \tag{4.3}$$

where $\theta_S$ is the *enob* of the front-end corresponding to sensor $S$, and $\beta_1$ and $\beta_2$ were defined in Sect. 3.2.1.

The objective function in line 13 of Algorithm 1 is defined as

$$OF(acc, cost) = \frac{\Delta acc}{\Delta cost},$$    (4.4)

where $\Delta[\cdot] = [\cdot]_t - [\cdot]_{t-1}$ at iteration $t$ and the cost is always normalized between 0 and 1, where 1 represents the configuration where all feature streams are extracted at the highest possible SNR. This normalization aims to ease the comparative performance analysis and addresses the lack of information on the properties of the circuit with which datasets were generated.

**Experimental Setup**

The performance of the ns-BN and the SCALEFEATURENOISE strategy was evaluated on 13 publicly available datasets from the UCI machine learning repository [4], as listed in Table 4.1. All the datasets underwent a pre-processing step consisting on the removal of nominal features and wrapper feature selection (with a naive Bayes classifier) to prevent over-fitting. Table 4.1 also provides the details of the datasets, including the number of features used for experiments (column denoted Selected F.).

For the experiments in this use case, the distribution $\Pr(F|C)$ in the ns-BN encodes a TAN classifier learned from data as explained in Sect. 2.2.3, and shown at the top of Fig. 4.2. The parameter vector for each sensor $S_i$, given in terms of SNR, was defined as: $\Theta_i = \{11, 10, 5, 4, 3, 2, 0\}$ *enob*, where 0 *enob* is equivalent to pruning all features in sensory stream $S_i$. This noise vector is selected to represent the performance of systems from recently published works on the topic [5].

**Table 4.1** Datasets for mixed-signal tuning experiments

| Dataset | Instances | Features | Classes | Selected F. |
|---|---|---|---|---|
| Pioneer[a] | 6129 | 27 | 35 | 17 |
| HAR-UCI | 10,299 | 561 | 6 | 37 |
| Sonar | 208 | 60 | 2 | 8 |
| Glass | 214 | 9 | 6 | 6 |
| WDBC | 569 | 30 | 2 | 11 |
| Pima | 768 | 8 | 2 | 5 |
| Vehicle | 846 | 18 | 4 | 8 |
| Banknote | 1372 | 5 | 2 | 5 |
| Leaf | 340 | 16 | 30 | 10 |
| Ecoli | 327 | 8 | 5 | 5 |
| Australian | 690 | 14 | 2 | 7 |
| Mhealth | 312,475 | 23 | 11 | 21 |
| PAMAP | 10,000[b] | 39 | 12 | 11 |

[a] From activities that include only linear movement (without turning and gripper activity)
[b] Reduced dataset size used for experiments

Finally, for all experiments, Algorithm 1 was initialized to the highest quality feature set ($\theta_{init} = \{11, \ldots, 11\}$ *enob*). The target accuracy and cost ($acc_{target}$, $cost_{target}$) were set to $\infty$ and 0, respectively, in such a way that the algorithm stops iterating when all features have been pruned and, as a consequence, the full trade-off space has been explored (locally). All the results presented throughout this section were generated by conducting 5-fold cross-validation and averaging the results over 5 trials (a new fold is generated for each trial).

## Results

The analysis herein focuses on the HAR [6] and Pioneer [7] datasets, since they correspond to applications that rely on sensors and are generally relevant for the topic of extreme-edge computing. The HAR dataset has 6 classes corresponding to the activities *Walking*, *Walking upstairs*, *Walking downstairs*, *Sitting*, *Standing*, and *Laying* and includes features extracted from a tri-axial accelerometer and a tri-axial gyroscope. The features used for experiments include mean, standard deviation, maximum/minimum magnitude, kurtosis, entropy, skewness, acceleration, and jerk, among others.

The Pioneer benchmark was collected by a mobile robot with three types of sensors (sonars, wheel odometers and vision sensors) and it has been used to identify 35 classes of navigation occurrences, described by the direction of the movement (forwards or backwards), whether the navigation path was obstructed or unobstructed, activity speed, and the visibility of objects in the environment.

Figure 4.4 shows the resulting local Pareto front in blue. The HAR bench-mark achieves cost savings of at least one order of magnitude without accuracy

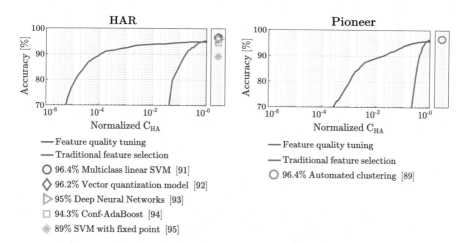

**Fig. 4.4** Pareto-optimal trade-off achieved by the noise tuning strategy on the HAR and on the Pioneer benchmarks compared to other benchmarks

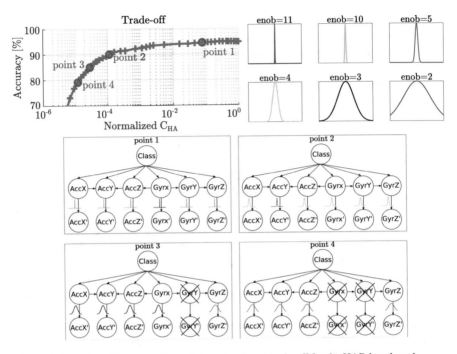

**Fig. 4.5** Examples of four operating points on the Pareto trade-off for the HAR benchmark

degradation. Additional cost savings of, for example, four orders of magnitude can be attained by selecting an operating point that performs at 90% instead of 95% accuracy. For the Pioneer benchmark, accuracy degradation is prevented with cost savings of an order of magnitude. But, in general, accuracy in the Pioneer benchmark degrades at a faster rate than HAR since the class variable of the former has a higher cardinality than the latter (35 versus 6 classes).

Figure 4.4 also includes a comparison with other approaches. The red curve in both sub-figures implements the method in [8], where features can be pruned by optimizing a resource-aware cost function. Furthermore, the figure shows the accuracy from other approaches found in literature. The HAR benchmark is compared with five other benchmarks [9–13] and the Pioneer benchmark with [7]. It is clear that the ns-BN achieves comparable baseline accuracy, and that it prevents this accuracy from degrading for a wide range of cost savings, unlike the simple feature pruning approach in red, which does not benefit from the mixed-signal quality scaling of the proposed approach.

Figure 4.5 illustrates the meaning of the local Pareto front in more detail for the HAR benchmark. Point 1 shows that cost savings of one order of magnitude without accuracy degradation can be achieved when the features generated by the x-gyroscope have an SNR of 11 *enob* while the rest are extracted with 10 *enob*. An additional cost reduction of three orders of magnitude (point 2)—with accuracy degradation of 5%—is achieved when the features from the y-accelerometer are

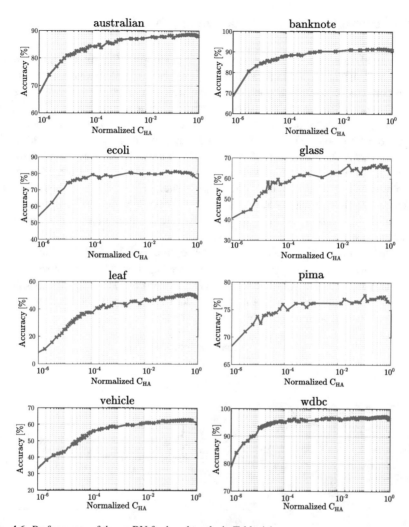

**Fig. 4.6** Performance of the ns-BN for benchmarks in Table 4.1

configured with an SNR of 5 *enob* and the rest with 4 *enob*. Points 3 and 4 are reached by pruning the features extracted from the y-gyroscope and the x-gyroscope, while allowing more noise on the rest of their features.

Finally, Fig. 4.6 demonstrates how the wide variety of benchmarks listed in Table 4.1 benefit from the proposed strategy: in all cases, cost savings of at least four orders or magnitude are procured with accuracy losses of less than 5%.

## 4.5   Second Use Case: Digital Quality Scaling

The second use case of the ns-BN targets the digital feature extraction block highlighted in Fig. 4.7a and seeks to exploit hardware implementations capable of tuning arithmetic operation precision at the front-end, like the one introduced in [14]. For the definition of the ns-BN in this use case, *noiseless* features **F** are those extracted at a nominal precision of $enob_{high}$, as specified by the capabilities of the hardware (the resolution of the ADC and the arithmetic precision supported by the digital hardware). Noisy feature versions **F'** are extracted at a lower precision of $enob_{low}$. For example, in Fig. 4.7b, $F_i$ in blue has a resolution of three bits, while $F_i'$ in red has a resolution of two bits. As mentioned in Sect. 3.3.1 for the cost model of Eq. (3.17), this use case assumes that the number of bits used in the digital domain is equal to the *enob* provided by the ADC, therefore representing a lower bound in digital precision.

To define the relation $Pr(F_i'|F_i)$ encoded by the ns-BN, the following observations are made: akin to the quantization error arising at an ADC,[7] the error induced by quantizing $F_i$ at resolution lower than the nominal is uniformly distributed between $-q_i/2$ and $q_i/2$, where $q_i$ is the quantization step of the lower resolution feature quantized with $enob_{low}$ bits and is given by

$$q_i = \frac{\max F_i - \min F_i}{2^{enob_{low}}}. \tag{4.5}$$

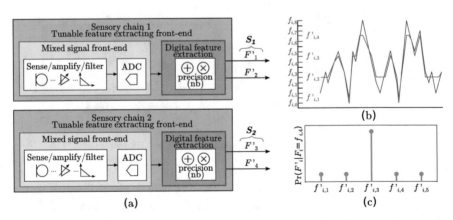

(a)

(b)

(c)

**Fig. 4.7** (**a**) Block diagram of the front-end with tunable-precision digital feature extraction. (**b**) Values of $F$ and $F'$. (**c**) Example of distribution $Pr(F_i'|F_i = f_{i,4})$

---

[7]Quantization error refers to the difference between an analog signal and the quantized version. The error in this use case considers the difference between high and low resolution digital signals and can be thus seen as *relative* quantization error.

In Fig. 4.7b, for example, $f'_{i,3}$ would have had a value equal to either $f_{i,4}$ or $f_{i,5}$ had it been quantized at a resolution of three bits instead of two.

Therefore, $\Pr(F'_i|F_i)$ is defined as a deterministic distribution: in the example above, $\Pr(F'_i|F_i = f_{i,4})$ and $\Pr(F'_i|F_i = f_{i,5})$ have a value of one at $F'_i = f'_{i,3}$ and zero elsewhere, whereas $\Pr(F'_i|F_i = f_{i,2})$ and $\Pr(F'_i|F_i = f_{i,3})$ are equal to one at $F'_i = f'_{i,2}$ (see Fig. 4.7c).

Analogous to the use case in Sect. 4.4, each sensor is interfaced to its own processing chain (see Fig. 4.7a). Each feature, or set of features from a given sensory chain, can be observed at $h$ different user defined precisions in number of bits $\Theta_i = \{enob_i^{(1)}, \ldots, enob_i^{(h)}\}$. According to the models discussed in Sect. 3.3, the energy consumption of digital systems can scale proportionally to the number of bits used for representation and arithmetic operations.

## 4.5.1 Experiments for Digital Quality Scaling

The experiments herewith attempt to identify the precision per sensory chain that leads to the local Pareto front.

### Cost and Objective Function Definition

As discussed in Sect. 3.6 and shown in Table 3.3, this use case assumes that the cost of digital feature extraction $C_F$ is predominant. The cost used for the evaluation of Algorithm 1 is based on Eq. (3.17) and motivated by Sect. 3.3.1:

$$\text{CostF}(\boldsymbol{\theta}) = \sum_{F \in \mathbf{F}} C_F(F) = \sum_{F \in \mathbf{F}} \beta \cdot \theta_F^2, \tag{4.6}$$

where $\theta_F$ is the number of bits for feature $F$ and $\beta$ is a feature-specific cost factor, determined according to the number of operations necessary to extract the feature. For the experiments herewith, it is assumed that all features are extracted, on average, with the same number of MAC operations, and therefore $\beta=1$. Note also that this model assumes that the quadratic scaling from the multiplication dominates with respect to the addition.

Finally, the objective function used in Algorithm 1 is given by

$$\text{OF}(acc, cost) = \log\left(\frac{\Delta acc}{\Delta cost}\right), \tag{4.7}$$

where cost is again normalized between 0 and 1. This calculation is performed in the logarithmic domain because, for some of the datasets considered in this section, accuracy has a considerably smaller dynamic range than cost, in contrast to the experiments in Sect. 4.4.1.

**Table 4.2** Datasets for digital feature tuning experiments

| Dataset | Instances | Features | Classes | Selected F. |
|---------|-----------|----------|---------|-------------|
| Synthetic | 2000 | 4 | 4 | 4 |
| HAD-USC | 208 | 60 | 4 | 8 |
| HAR-RIO | 214 | 9 | 6 | 6 |
| Robot Nav. | 10,299 | 561 | 6 | 37 |

**Experimental Setup**

Table 4.2 describes the four datasets used for the experiments in this section. The first one is a synthetic dataset generated by sampling from multivariate Gaussian distributions described in Appendix A.1. The second and third relate to activity recognition applications: the HAD-USC [15] dataset includes features extracted from tri-axial accelerometers and gyroscopes and considers four types of activities, and HAR-RIO [16] includes five activities and features extracted with four accelerometers. Finally, the Robot Nav. dataset [17] was collected by a mobile robot equipped with ultrasound sensors and considers six types of navigation activities.

For all the experiments in this section, there were $h = 5$ possible settings per feature, given in number of bits. Motivated by the implementations discussed in [14], the values $\Theta_i = \{10, 8, 5, 4, 3, 0\}$ were selected for all features, with 0 corresponding to pruning feature $i$. Similar to the experiments in Sect. 4.4.1, Algorithm 1 was initialized on the highest quality feature set and was executed until all features were pruned. Finally, the distribution $\Pr(\mathbf{F}|C)$ for the ns-BN in these experiments represents a naive Bayes classifier (see the middle of Fig. 4.2), with parameters learned from training data with Maximum-Likelihood estimation [18].

Due to the size of some of the datasets available, experiments were not cross-fold validated in this section. Rather, they were evaluated on random 75–25% train-test splits with five trials, each with a different split. Moreover, training data was used to evaluate the accuracy of Algorithm 1, i.e. $\mathcal{F}_{valid}$ was the training dataset.

**Results**

Figure 4.8 shows in red the resulting Pareto accuracy versus normalized cost. The blue line shows the results of cost-aware feature selection strategy, where the cost function in Eq. (4.7) is used to decide whether features are observed or pruned.

The proposed approach proves to be effective in all datasets, with extensive cost-saving possibilities, beyond those enabled by traditional binary feature selection. In three of the four datasets, the proposed method is capable of preserving the initial accuracy with cost reductions of more than 85%, while the traditional feature selection techniques see a faster degradation of accuracy. Moreover, the ns-BN based method results in a finer grained Pareto-optimal front, in such a way that the user can select feature configurations that more precisely meet accuracy and cost specifications.

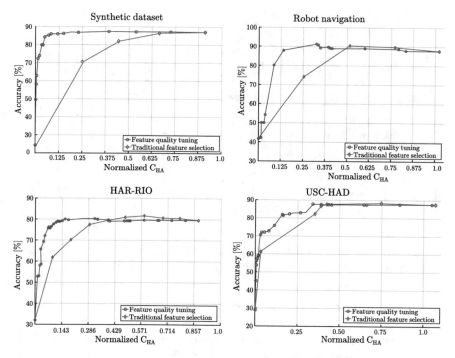

**Fig. 4.8** Experimental results for the digital quality scaling use case

## 4.6   Third Use Case: Analog Quality Scaling

The final use case of the ns-BN targets applications that extract features in the analog domain. The guiding principle of this hardware design approach is to perform early information discrimination by extracting features as close as possible to the sensor, forgoing the need to discretize all incoming data.

Figure 4.9a shows an example of this type of sensor front-end: analog features are extracted by decomposing the incoming sensory signal into bands of varying frequency. Each band is then completed by an ADC with a resolution of *enob*. In addition to deciding whether to extract a feature or not, the tunability aspect in this system is determined by how effectively the ADC's quantization levels are used. Each feature extracting band is equipped with an amplifier, which can be tuned to determine the *enob* at the output of that particular band: because the ADC of each band is configured to amplify a signal with a given peak to peak amplitude, reduced signal amplitude can result in lower utilization of the available quantization levels, as illustrated by Fig. 4.9b. This example shows a 3 bit ADC, with 8 quantization levels available. The blue signal, with a larger peak to peak amplitude, uses all available quantization levels effectively, while the red one only uses two.

The quality of the output feature at each band is therefore expressed in terms of the effectively utilized bits (or *enob*). In this use case, the *noiseless*—or high

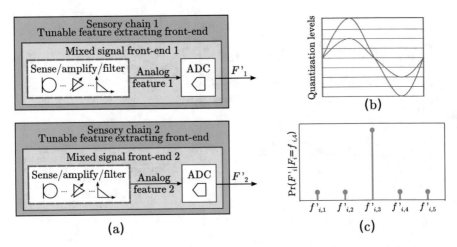

**Fig. 4.9** (**a**) Feature extraction in the analog domain. (**b**) Example of the impact of signal range on quantization level usage. (**c**) Distribution of noisy features given noiseless features

resolution—features $F_i$ in the ns-BN are those quantized with the highest possible resolution. In other words, these high resolution features are extracted with the maximum number of bits procured by the ADC. Furthermore, the probabilistic relation $\Pr(F_i'|F_i)$ can also be encoded by a deterministic distribution. This use case therefore follows the same approach as the use case in Sect. 4.5 in terms of modeling and inference.[8]

### 4.6.1  Experiments for Analog Quality Scaling

The experiments in this section consider the voice activity detection (VAD) application which consists on identifying whether a given sound corresponds to voice or not. More specifically, the properties of the VAD front-end introduced in [2] guide the definition of the model and the hardware-aware cost. This system extracts features in the analog domain by decomposing the incoming audio signal into frequency bands. Figure 4.10b shows an example of one of the sixteen bands: the amplified signal from a passive microphone is fed to a Band Pass Filter (BPF), whose output is then rectified and averaged to obtain the analog feature. The center frequency of the Band Pass Filter (BPF) increases exponentially from 75 Hz to 5 kHz from band 1 to band 16, and the ADC at the output of each band can discretize the signal at 8 bits. Further, for the experiments in this section, it is assumed each band

---

[8]Note that $\Pr(F_i'|F_i)$ could also be modeled as a Gaussian distribution, like in the first use case in Sect. 4.4.

**Fig. 4.10** (a) Block diagram for VAD. (b) Variable precision feature extraction for VAD. Based on [19]

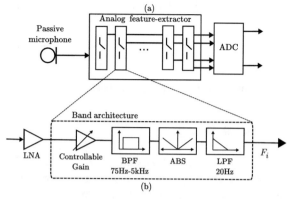

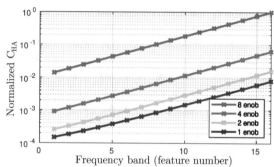

**Fig. 4.11** Energy consumption scaling for the VAD system: it scales exponentially from band 1 to 16 and exponentially with each additional bit

has variable amplification for precision-tunability such that each band is capable of extracting each feature at 8, 4, 2, and 1 *enob* [19].

**Cost and Objective Function Definition**

This use case relies in the analog front-end block for feature extraction. As explained in Sect. 3.3.2, the cost associated with extracting features in the analog domain follows the same trends as the tunable mixed-signal front-end studied in Sect. 4.4. Therefore, the hardware-aware cost function for the experiments in based in Eq. (3.18):

$$\text{CostF}(\boldsymbol{\theta}) = \sum_{F \in \mathbf{F}} C_{F,\text{analog}}(F) = \sum_{F \in \mathbf{F}} \beta \cdot 2^{\theta_F}, \qquad (4.8)$$

where $\theta_F$ is the *enob* of feature $F$, as described above. The value of $\beta$ is based on the energy consumption measurements of the VAD chip proposed in [2] and shown in Fig. 4.11. In this system, energy consumption scales exponentially from feature 1 to 16 and also exponentially with each additional bit, similar to the energy scaling from Eq. (4.3).

The objective function used for Algorithm 1 is the same as in the previous use case:

$$OF(acc, cost) = \log\left(\frac{\Delta accuracy}{\Delta cost}\right), \qquad (4.9)$$

where cost is again normalized between 0 and 1.

**Experimental Setup**

Experiments in this section were performed on the NOIZEUS corpus [20], which consists of IEEE sentences spoken by three male and three female speakers, subjected to several types of environmental noises, including *suburban train, babble, car, exhibition hall, restaurant, street, airport,* and *train station.*

The experiments herewith focused on the *restaurant, suburban train, exhibition,* and *subway* contexts as they are representative of the environmental conditions an extreme-edge VAD is likely to encounter. Furthermore, these instances of environmental noise are assumed to manifest at three different SNR levels (3 db, 9 dB, 15 dB). Accuracy was evaluated on the same 50% train-set random split scheme from the experiments in [2]. Finally, one ns-BN was trained per context type.

In terms of the ns-BN for these experiments, the probability $Pr(\mathbf{F}|C)$ represents a naive Bayes classifier (see the bottom of Fig. 4.2), and the configuration vector for features $F'i$ was defined as $\Theta_i = \{8, 4, 2, 1, 0\}enob$ , where 0 refers to pruning $F_i$.

**Results**

Figure 4.12 shows the results on the VAD application under *suburban train* context at three SNR levels. The proposed strategy achieves cost savings of more than three orders of magnitude with accuracy degradation of less than 5% in all three cases.

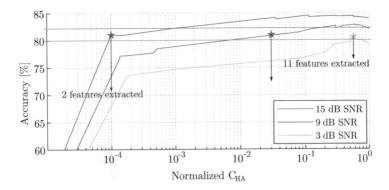

**Fig. 4.12** Cost versus accuracy trade-off for the *suburban train* context under different SNR levels

**Table 4.3** Selected feature sets for the *suburban train* context at three noise levels

| Feature $F$ | $\theta_F$ [enob] | | |
|---|---|---|---|
| | 3 dB SNR | 9 dB SNR | 15 dB SNR |
| 1 to 5 | – | – | – |
| 6 | 1 | – | – |
| 7 | 4 | 4 | 4 |
| 8 | 8 | 8 | 8 |
| 9 and 10 | 8 | – | – |
| 11 | 1 | – | – |
| 12 | 8 | 1 | – |
| 13 | 2 | – | – |
| 14 and 15 | 8 | – | – |
| 16 | 8 | 1 | – |
| Normalized $C_{HA}$ | 0.6 | 0.03 | 0.0001 |
| Accuracy [%] | 80 | 81 | 81 |

The markers illustrate a scenario where the user requests that accuracy is between 80% and 82% (as bounded by the green dotted lines). This is achieved with a very low cost when environmental conditions are favorable (15 dB SNR) by relying on only two features, as also shown in Table 4.3. When environmental conditions are more challenging (3 dB SNR), 11 features must be extracted to meet the desired accuracy at a higher cost.

Figure 4.13 compares the cost versus accuracy trade-off of the VAD application under four contexts at 3 dB SNR. Note that contexts that involve speech in their background noise, such as *restaurant* and *subway*, are more challenging for the VAD classifier, resulting in lower accuracy overall. The *suburban train* context sees a steady decrease in accuracy as features are pruned and represented with lower precision (see also Table 4.3). In other contexts, accuracy remains steady or increases when pruning features. This could be due to over-fitting. But also due to the fact that some of the features share the same energy bands of background noise. These features, therefore, tend to classify background noise as voice, and removing them improves the performance of the classifier.

## 4.7   Related Work

From a hardware design standpoint, many state-of-the-art implementations focus on the optimization and design of energy efficient hardware accelerators that exploit stochastic or low-precision computing [21, 22]. Others focus on efficient sensor interfaces, motivated by energy consumption dominance in that block, which therefore constitutes the main bottleneck towards resource-efficient operation. For systems that do not require full precision, but need always-on sensing, adaptive sensing based approaches have been proposed [2, 23]. Some of the techniques

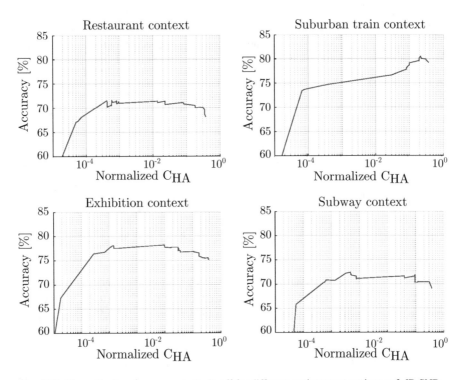

**Fig. 4.13** Normalized cost vs accuracy trade-off for different environment settings at 3 dB SNR

presented throughout this chapter can focus on the exploitation of these hardware-scalable works.

From the resource-aware machine learning point of view, one common approach is to sequentially select the set of observations that provide the most information [24]. Another approach consists on determining whether more observations are required to meet the tasks' requirements [25]. Both of these approaches can be incorporated to the noise-scaling algorithm introduced in this chapter.

Another relevant body of work focuses on feature subset selection, where there is a trade-off between the observations' value of information and the impact these observations have on the overall system's resource management, whether computational or circuit oriented [8, 13, 26]. A third line of research exploits application specific temporal characteristics, such as context changes, to optimally switch to low-power sensing modalities that keep inference performance losses to a minimum [27]. Finally, the authors of [28] have devised a run-time power-vs-quality tuning strategy for sensor networks that relies on a cascaded detection system to filter out irrelevant data instances and thus reduces data transmission costs while also increasing overall feature quality. However, the two aforementioned techniques exploit application specific properties, context changes in the first case, and data transmission in the second.

In addition, all the techniques mentioned thus far only enable a few points in the power versus accuracy trade-off space, as they can only decide to observe a feature or not. This fails to exploit all the cost-saving opportunities of the hardware platform, which can, in some instances, be tuned to operate at different levels of quality and can be exploited by the methods introduced in this chapter.

## 4.8  Discussion

This chapter proposed the noise-scalable Bayesian Network classifier, a model that can encode hardware-induced feature quality degradation and that allows to evaluate classification accuracy of feature sets extracted at different levels of quality. This model can be used to search the hardware-aware cost versus accuracy trade-off space efficiently, allowing to derive the feature-wise noise configuration that constitute the Pareto-optimal front within the reachable space. The versatility of this model was proven by evaluating its performance in three use cases, each exploiting different scalable stages of the feature extraction pipeline and different types of hardware-induced noise. Experiments show that the proposed model and search strategy achieve significant cost savings with limited accuracy losses. Furthermore, they extract a fine grained local Pareto feature-configuration set that can be used to determine the amount of noise permitted in a sensor front-end in accordance to performance requirements. This can be used to set the run-time configuration of the device that will perform at the desired accuracy and cost. This model can also be used to re-configure the front-end's quality at run-time, as will be discussed in Chap. 6.

The models and methods proposed throughout this section make a number of assumptions. The most notable are: (1) The fact that the noise arising at every sensory chain is assumed to be uncorrelated to the noise of the other chains. (2) The fact that features extracted from the same sensory chain remain independent from each other, and that all features, regardless of their sensory chain, experience the same type and amount of noise. (3) In experiments, all features have the same levels of noise available, which might not be the case in all applications.

These considerations should motivate future work, and especially, the practical implementation of the methods proposed in this chapter. Finally, it should also be taken into consideration that the hardware-aware cost derived in Chap. 3 was based on the purely physical limits of circuits at the front-end and that they made simplifying assumptions on the components included therein. As also mentioned in Sect. 3.2.1, these cost models may not hold in practice for low SNR values, where most of the cost-saving opportunities are found for the methods proposed in this chapter. However, the proposed model and strategy can be easily updated to account for more complex and realistic models, as was proven from the study of three different use cases and the showcase of their functionality on a variety of datasets.

# References

1. L. Galindez, K. Badami, J. Vlasselaer, W. Meert, M. Verhelst, Dynamic sensor-frontend tuning for resource efficient embedded classification. IEEE J. Emerg. Select. Top. Circ. Syst. **8**(4), 858–872 (2018)
2. K.M. Badami, S. Lauwereins, W. Meert, M. Verhelst, A 90 nm CMOS, power-proportional acoustic sensing frontend for voice activity detection. IEEE J. Solid-State Circ. **51**(1), 291–302 (2016)
3. M.J. Pelgrom, Analog-to-digital conversion, in *Analog-to-Digital Conversion* (Springer, New York, 2013), pp. 325–418
4. M. Lichman, UCI machine learning repository, 2013 [Online]. Available: http://archive.ics.uci.edu/ml
5. J. De Roose, H. Xin, M. Andraud, P.J. Harpe, M. Verhelst, Flexible and self-adaptive sense-and-compress for sub-microwatt always-on sensory recording, in *ESSCIRC 2018-IEEE 44th European Solid State Circuits Conference (ESSCIRC)* (IEEE, New York, 2018), pp. 282–285
6. D. Anguita, A. Ghio, L. Oneto, X. Parra, J.L. Reyes-Ortiz, A public domain dataset for human activity recognition using smartphones, in *ESANN*, 2013
7. T. Oates, M.D. Schmill, P.R. Cohen, Identifying qualitatively different outcomes of actions: gaining autonomy through learning, in *AGENTS '00*, 2000
8. H. Ghasemzadeh, N. Amini, R. Saeedi, M. Sarrafzadeh, Power-aware computing in wearable sensor networks: an optimal feature selection. IEEE Trans. Mobile Comput. **14**(4), 800–812 (2015)
9. B. Romera-Paredes, M.S. Aung, N. Bianchi-Berthouze, A one-vs-one classifier ensemble with majority voting for activity recognition, in *ESANN*, 2013, pp. 443–448
10. M. Kästner, M. Strickert, T. Villmann, S.-G. Mittweida, A sparse kernelized matrix learning vector quantization model for human activity recognition, in *ESANN*, 2013
11. C.A. Ronao, S.-B. Cho, Human activity recognition with smartphone sensors using deep learning neural networks. Expert Syst. Appl. **59**, 235–244 (2016)
12. A. Reiss, G. Hendeby, D. Stricker, A competitive approach for human activity recognition on smartphones, in *European Symposium on Artificial Neural Networks, Computational Intelligence and Machine Learning (ESANN 2013), 24–26 April, Bruges* (ESANN, Bruges, 2013), pp. 455–460
13. D. Anguita, A. Ghio, L. Oneto, X. Parra, J.L. Reyes-Ortiz, Human activity recognition on smartphones using a multiclass hardware-friendly support vector machine, in *International Workshop on Ambient Assisted Living* (Springer, New York, 2012), pp. 216–223
14. B. Moons, M. Verhelst, DVAS: dynamic voltage accuracy scaling for increased energy-efficiency in approximate computing, in *2015 IEEE/ACM International Symposium on Low Power Electronics and Design (ISLPED)* (IEEE, New York, 2015), pp. 237–242
15. M. Zhang, A.A. Sawchuk, USC-HAD: a daily activity dataset for ubiquitous activity recognition using wearable sensors, in *ACM International Conference on Ubiquitous Computing (Ubicomp) Workshop on Situation, Activity and Goal Awareness (SAGAware)*, Pittsburgh, PA, September 2012
16. W. Ugulino, D. Cardador, K. Vega, E. Velloso, R. Milidiú, H. Fuks, Wearable computing: accelerometers' data classification of body postures and movements, in *Advances in Artificial Intelligence-SBIA 2012* (Springer, New York, 2012), pp. 52–61
17. A.L. Freire, G.A. Barreto, M. Veloso, A.T. Varela, Short-term memory mechanisms in neural network learning of robot navigation tasks: a case study, in *2009 6th Latin American Robotics Symposium (LARS)*, October 2009, pp. 1–6
18. A. Darwiche, *Modeling and Reasoning with Bayesian Networks* (Cambridge University Press, Cambridge, 2009)
19. L.I. Galindez Olascoaga, K. Badami, V.R. Pamula, S. Lauwereins, W. Meert, M. Verhelst, Exploiting system configurability towards dynamic accuracy-power trade-offs in sensor front-ends, in *2016 50th Asilomar Conference on Signals, Systems and Computers* (IEEE, New York, 2016), pp. 1027–1031

20. Y. Hu, Subjective evaluation and comparison of speech enhancement algorithms. Speech Commun. **49**, 588–601 (2007)
21. B. Moons, M. Verhelst, Energy-efficiency and accuracy of stochastic computing circuits in emerging technologies. IEEE J. Emerg. Select. Top. Circ. Syst. **4**(4), 475–486 (2014)
22. E. Park, D. Kim, S. Yoo, Energy-efficient neural network accelerator based on outlier-aware low-precision computation, in *2018 ACM/IEEE 45th Annual International Symposium on Computer Architecture (ISCA)* (IEEE, New York, 2018), pp. 688–698
23. M. Yip, A.P. Chandrakasan, A resolution-reconfigurable 5-to-10-bit 0.4-to-1 V power scalable SAR ADC for sensor applications. IEEE J. Solid-State Circ. **48**(6), 1453–1464 (2013)
24. A. Krause, C.E. Guestrin, Near-optimal nonmyopic value of information in graphical models (2012). Preprint. arXiv:1207.1394
25. A. Choi, Y. Xue, A. Darwiche, Same-decision probability: a confidence measure for threshold-based decisions. Int. J. Approx. Reason. **53**(9), 1415–1428 (2012)
26. S. Lauwereins, K. Badami, W. Meert, M. Verhelst, Context and cost-aware feature selection in ultra-low-power sensor interfaces, in *European Symposium on Artificial Neural Networks, Computational Intelligence and Machine Learning*, 2014, pp. 93–98
27. D. Jun, L. Le, D.L. Jones, Cheap noisy sensors can improve activity monitoring under stringent energy constraints, in *2013 IEEE Global Conference on Signal and Information Processing*, December 2013, pp. 683–686
28. L.N. Le, D.L. Jones, Guided-processing outperforms duty-cycling for energy-efficient systems. IEEE Trans. Circ. Syst. I: Regul. Pap. **64**(9), 2414–2426 (2017)

# Chapter 5
# Hardware-Aware Probabilistic Circuits

Chapter 4 proposed a hardware-aware probabilistic model that targets the feature extracting block of the embedded sensing pipeline. This realization of hardware-awareness owes to two main enabling factors of Bayesian networks: (1) They allow to encode domain knowledge, both through their structure and through the parametrization of their associated probability distributions, which has been leveraged to encode the hardware-driven scalable signal quality. Furthermore, they can complement these knowledge encoding capabilities with the benefits of data-driven parameter learning. (2) Bayesian networks encode joint probability distributions over a set of random variables and are equipped to handle marginal queries over any subset of those random variables. This is key in encoding the different sensor streams available in many embedded applications and in determining the impact that disabling features and sensors can have on the cost versus accuracy trade-off.

As discussed in Sect. 3.6, this chapter considers the costs incurred by the full sensory embedded pipeline. Therefore, the strategies proposed herewith look beyond the feature extracting front-end block for additional cost-saving opportunities available in the system, such as those offered by scaling inference complexity. This is achieved with a hardware-aware Probabilistic Circuit (PC)[1] that considers scalable inference complexity in addition to scalable feature extraction. This system-wide hardware-awareness is enabled by PCs' properties: (1) Learning these models explicitly trades-off model fitness and size. This learning goal is brought a step further by taking into consideration the hardware-aware costs of the (incrementally) learned models. (2) As computational graphs, PCs provide a fixed formula for the exact calculation of marginal probabilities. This is used for the straightforward calculation of the system-wide hardware-aware costs, as discussed in Sect. 3.4. This cost calculation is then used by the search strategy proposed in this chapter,

---

[1] As discussed in Sect. 2.3, PGMs can also be converted into PCs by a process known as knowledge compilation [1]. PGMs compiled to arithmetic circuits will be used in this book mainly for benchmarking purposes.

© The Author(s), under exclusive license to Springer Nature Switzerland AG 2021
L. I. Galindez Olascoaga et al., *Hardware-Aware Probabilistic Machine Learning Models*,
https://doi.org/10.1007/978-3-030-74042-9_5

which outputs the system-wide configurations (in terms of sensor and feature quality as well as model complexity) that constitute the locally optimal Pareto accuracy versus cost trade-off. (3) Like Bayesian networks, PCs represent joint probability distributions, a property that is leveraged to measure the effects of sensor and feature pruning on accuracy. This robustness is also procured by the PCs in this chapter, but classification accuracy is further improved by leveraging these models' ability to encode logical constraints.

This chapter begins in Sect. 5.1 with an introduction to the properties of Probabilistic Sentential Decision Diagrams (PSDDs), the type of PC relied upon by the contributions to follow. Section 5.2 formalizes the proposed resource-aware cost metric that takes into consideration the hardware's properties in determining whether the inference task can be efficiently deployed. A multi-stage trade-off space search strategy based on the aforementioned cost metric is then introduced in Sect. 5.3, and Sect. 5.4 empirically evaluates it on several publicly available datasets. The work presented in these first five sections is based on the following publication:

Galindez Olascoaga, L. I., Meert, W., Shah, N., Verhelst, M. and Van den Broeck, G. (2019). Towards Hardware-Aware Tractable Learning of Probabilistic Models. *In Advances in Neural Information Processing Systems (NeurIPS)* (pp. 13726–13736).

Section 5.5 then introduces a technique that encodes a discriminative bias within a generative PC learning algorithm, and Sect. 5.6 verifies this method with a number of experiments on classification benchmarking datasets. This technique was previously published in:

Galindez Olascoaga, L. I., Meert, W., Shah, N., Van den Broeck, G., and Verhelst, M. (2020). Discriminative Bias for Learning Probabilistic Sentential Decision Diagrams. *In International Symposium on Intelligent Data Analysis (IDA)* (pp. 184–196). Springer, Cham.

Finally, Sects. 5.7 and 5.8 discuss related work and offer conclusions, respectively.

## 5.1   Preliminaries

The work introduced in this chapter relies mainly on the Probabilistic Sentential Decision Diagram (PSDD), a type of PC that can represent a joint distribution over a set of random variables and whose structure can be learned in a generative fashion and fully from data. This section gives an overview of the properties of this model and introduces the learning technique exploited by the strategies introduced in this chapter.

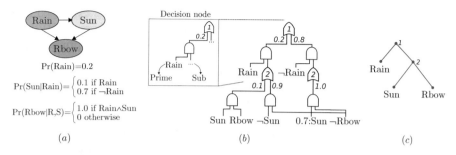

**Fig. 5.1** A Bayesian network and its equivalent PSDD (adapted from [3]). (**a**) Bayesian network and its CPTs. (**b**) Equivalent PSDD. (**c**) PSDD's vtree

## 5.1.1 Probabilistic Sentential Decision Diagrams

PSDDs were introduced as probabilistic extensions to Sentential Decision Diagrams (SDDs) [2], which represent Boolean functions as logical circuits. The inner nodes of a PSDD alternate between AND gates—equivalent to product nodes—with two inputs and OR gates—equivalent to sum nodes—with arbitrary number of inputs; the root must be an OR node; and each leaf node encodes a distribution over a variable $X$ (see Fig. 5.1c). The combination of an OR gate with its AND gate inputs is referred to as *decision* node, where the left input of the AND gate is called *prime* ($p$), and the right is called *sub* ($s$). Each of the $n$ edges of a decision node is annotated with parameters $\theta_1, \ldots, \theta_n$, where $\sum_i \theta_i = 1$.

PSDDs are subjected to three structural constraints: decomposability, smoothness, and determinism, which were defined in Sect. 2.3.2. Decomposability is enforced by a *vtree*, a binary tree whose leaves are the random variables and that determines how variables are arranged in primes and subs in the PSDD (see Fig. 5.1c): each internal vtree node is associated with the PSDD nodes at the same level, variables appearing in the left subtree $\mathbf{X}$ are the primes, and the ones appearing in the right subtree $\mathbf{Y}$ are the subs. Determinism means that only one of the inputs (children) of each decision node can be true as shown in Fig. 5.1b: *Rain* can either be true with a probability of 0.8 or false with a probability of 0.2.

Each PSDD node $q$ represents a probability distribution. Terminal nodes encode univariate distributions. Decision nodes, when normalized for a vtree node with $\mathbf{X}$ in its left subtree and $\mathbf{Y}$ in its right subtree, encode the following distribution over $\mathbf{XY}$ (see also Fig. 5.1a and b):

$$\text{Pr}_q(\mathbf{XY}) = \sum_i \theta_i \text{Pr}_{p_i}(\mathbf{X}) \text{Pr}_{s_i}(\mathbf{Y}). \tag{5.1}$$

Thus, each decision node decomposes the distribution into independent distributions over $\mathbf{X}$ and $\mathbf{Y}$. In general, prime and sub variables are independent at PSDD node $q$ given the prime *base* $[q]$ [4]. This base is the support of the distribution encoded learning task, unlike the *enob* used throughout the previous chapter, determining the quality of feature extraction.

## 5.3  Pareto-Optimal Trade-off Extraction

It is clear that the $C_{HA}$ of the systems considered in this chapter depends on four system properties: (1) the complexity of model $\kappa$, determined by the number and type of its operations; (2) the size and type of the feature set $\mathbf{F}$; (3) the size and type of the available sensor set $\mathbf{S}$; and (4) the number of bits $nb$ used within $\kappa$. An instantiation of these four properties is hereby referred to as a *system configuration* and is denoted by $\zeta = \{\kappa, \mathbf{F}, \mathbf{S}, nb\}$.

Clearly, the system configuration also determines the model's classification accuracy. Recall from Sect. 2.3.3 that classification with PCs can be performed with Bayes rule after computing the relevant marginal probabilities. As mentioned before, the goal of the strategy proposed in this section is to select the system configurations that map to the (local) Pareto-frontier on the hardware-cost versus accuracy trade-off space. The inputs to the strategy are the class variable $C$, the available features $\mathbf{F}$ and sensors $\mathbf{S}$ sets, and the set of available precisions $\mathbf{nb}$. The output is the set of Pareto system configurations $\zeta^* = \{\{\kappa_i^*, \mathbf{F}^*_i, \mathbf{S}^*_i, nb_i^*\}_{i=1:p}\}$.

The proposed strategy searches the cost versus accuracy trade-off space by sequentially scaling the four system properties:

**Model Complexity Scaling**  A set of PCs $\kappa$ of increasing complexity is first learned on training data $\mathcal{F}_{\text{train}}$. Each maps to a specific classification accuracy and inference cost $C_M$ (see Eq. (5.5)). This step relies on the LEARNPSDD algorithm [3] for two main reasons: this algorithm improves the model incrementally, but each iteration already leads to a valid PC that can be used to populate the set $\kappa$. Moreover, as discussed in Sect. 5.1, each of the learned PSDDs encodes a joint probability distribution. Like in the case of Bayesian networks, this allows them to remain robust to missing values, either due to malfunction as will be illustrated in Sect. 5.6 or to enforce cost efficiency as illustrated throughout Chap. 4 when features and sensors were pruned.

**Feature and Sensor Set Scaling**  The feature set $\mathbf{F}$ is scaled by sequentially pruning individual features. The impact on feature cost $C_F$ is clear from the discussions in Sect. 3.3, but pruning features can also have an impact on the inference costs $C_M$: if a Boolean random variable always remains unobserved, its corresponding indicator variables remain fixed (see, for example Fig. 2.3), which is exploited by Algorithm 4 to simplify the circuit. In addition, sensor $S \in \mathbf{S}$ can be pruned when none of the features it originates is used anymore.

**Precision Scaling**  Reducing the precision of arithmetic operations and numerical representations entails information loss and results in performance degradation [7]. The effect on inference costs $C_M$ is clear from Eq. (5.5). The experiments in this

---

**Algorithm 3:** SCALESI($\kappa_l$, $\mathbf{F}_{prun}$, $\mathbf{S}_{prun}$)

---

**Input:** $\kappa_l$: the $l$th model in $\kappa$, $\mathbf{F}_{prun}$, $\mathbf{S}_{prun}$: set of prunable features and sensors.

**Output:** $\langle \mathcal{F}^l, \mathcal{S}^l, \mathcal{M}^l \rangle$, **acc**, **cost**: $l$th collection of pruned features, sensors and model sets, their accuracy and cost.

1  $\mathbf{F}_{sel} \leftarrow \mathbf{F}_{prun}$, $\mathbf{S}_{sel} \leftarrow \mathbf{S}_{prun}$, $\kappa_{sel} \leftarrow \kappa_l$
2  $\langle \mathcal{F}^l, \mathcal{S}^l, \mathcal{M}^l \rangle \leftarrow \langle \mathbf{F}_{sel}, \mathbf{S}_{sel}, \kappa_{sel} \rangle$
3  $acc_{sel}$=Acc($\alpha_{sel}$, $\mathbf{F}_{sel}$), $cost_{sel}$=CHA($\alpha_{sel}$, $\mathbf{F}_{sel}$, $\mathbf{S}_{sel}$)
4  $\langle$**acc**, **cost**$\rangle \leftarrow \langle acc_{sel}, cost_{sel} \rangle$
5  **while** $|\mathbf{F}_{sel}| > 1$ **do**
6  $\quad$ $ob_{max} \leftarrow$                          `// Initialize objective value`
7  $\quad$ **foreach** $F \in \mathbf{F}_{prun}$ **do**
8  $\quad\quad$ $\mathbf{F}_{ca} \leftarrow \mathbf{F}_{sel} \setminus F$
9  $\quad\quad$ $\mathbf{S}_{ca} \leftarrow \mathbf{S}_{sel}$
10 $\quad\quad$ **foreach** $S \in \mathbf{S}_{prun}$ **do**
11 $\quad\quad\quad$ **if** $\mathbf{F}_{ca} \cap \mathbf{F}_S = \varnothing$ **then**
12 $\quad\quad\quad\quad$ $\mathbf{S}_{ca} \leftarrow \mathbf{S}_{ca} \setminus S$                 `// Prune sensor`
13 $\quad\quad$ $\kappa_{ca} \leftarrow$PrunePC($\kappa_{sel}$, $\mathbf{F}_{ca}$)
14 $\quad\quad$ $acc_{ca} \leftarrow$ Acc($\alpha_{ca}$, $\mathbf{F}_{ca}$)
15 $\quad\quad$ $cost_{ca} \leftarrow$ CHA($\alpha_{ca}$, $\mathbf{F}_{ca}$, $\mathbf{S}_{ca}$)
16 $\quad\quad$ $ob_{ca} \leftarrow$ OF($acc_{ca}$, $cost_{ca}$)
17 $\quad\quad$ **if** $ob_{ca} > ob_{max}$ **then**
18 $\quad\quad\quad$ $ob_{max} \leftarrow ob_{ca}$
19 $\quad\quad\quad$ $\mathbf{F}_{sel} \leftarrow \mathbf{F}_{ca}$, $\mathbf{S}_{sel} \leftarrow \mathbf{S}_{ca}$, $\kappa_{sel} \leftarrow \kappa_{ca}$
20 $\quad\quad\quad$ $acc_{sel} \leftarrow acc_{ca}$, $cost_{sel} \leftarrow cost_{ca}$
21 $\quad$ $\mathcal{F}^l$.insert($\mathbf{F}_{sel}$), $\mathcal{S}^l$.insert($\mathbf{S}_{sel}$), $\mathcal{M}^l$.insert($\kappa_{sel}$)
22 $\quad$ **acc**.insert($acc_{sel}$), **cost**.insert($cost_{sel}$)
23 **return** $\langle \mathcal{F}^l, \mathcal{S}^l, \mathcal{M}^l \rangle$, **acc**, **cost**

---

chapter will consider the standard IEEE 754 floating point representations, as they can be implemented in most embedded hardware platform.

## 5.3.1   Search Strategy

Finding the smallest possible PC that computes a given function is $\Sigma_2^p$-hard [8], thus computationally harder than NP. No single optimal solution is known for this problem; it is a central question in the field of knowledge compilation [1]. Optimizing for the lowest cost/highest accuracy PC further increases complexity. Therefore, this chapter again opts for a greedy optimization strategy that relies on a series of heuristics to search the trade-off space. Each step independently scales one of the configuration properties $\langle \kappa, \mathbf{F}, \mathbf{S}, nb \rangle$, as described in the previous section, and aims to find its locally optimal setting. The search begins by learning the model set $\kappa = \{\kappa_l\}_{l=1:n}$. Then, starting from each model $\kappa_l$, Algorithm 3 executes a greedy neighborhood search that maximizes cost savings and minimizes accuracy losses by

---

**Algorithm 4:** PRUNEPC($\kappa$,**F**)

---

**Input**: $\kappa$: the input PC, **F**: the observed feature set used to guide the pruning of $\kappa$.
**Output**: $\kappa_{pr}$: the pruned AC.

1　$\kappa_{pr} \leftarrow \text{copy}(\kappa)$
　　/* Loop through PC, children before parents　　　　　　　　　　*/
2　**foreach** $k$ *in* $\kappa_{pr}$ **do**
3　　　**if** $k$ *is an indicator variable* $\lambda_{F=f}$ *and* $F \notin \mathbf{F}$ **then**
4　　　　　replace $k$ in $\kappa_{pr}$ by a constant 1
5　　　**else if** $k$ *is* $+$ *or* $\times$ *with constant children* **then**
6　　　　　replace $k$ in $\kappa_{pr}$ by an equivalent constant

7　**return** $\kappa_{pr}$

---

sequentially pruning the sets **F** and **S** and simplifying $\kappa_l$ accordingly (Algorithm 4). Each iteration evaluates the accuracy and cost of $m$ feature subset candidates, where each considers the impact of removing a feature from the user defined prunable set $\mathbf{F}_{prun} \subseteq \mathbf{F}$. The algorithm then selects the feature and sensor subsets $\mathbf{F}_{sel} \subseteq \mathbf{F}$, $\mathbf{S}_{sel} \subseteq \mathbf{S}$ and the simplified model $\kappa_{sel}$ that maximizes the objective function $OF = acc/cost_{norm}$, where $cost_{norm}$ is the evaluated hardware-aware cost $C_{HA}$, normalized according to the maximum achievable cost (from the most complex model $\kappa_n$). Note that feature subset selection drives sensor subset selection $\mathbf{S}_{sel}$, as described before and defined in lines 12 and 19 of Algorithm 3.

The output of Algorithm 3, $\langle \mathcal{F}^{(l)}, \mathcal{S}^{(l)}, \mathcal{M}^{(l)} \rangle$, is a set of system configurations of the form $\{\{\mathbf{F}_{sel,1}, \mathbf{S}_{sel,1}, \kappa_{sel,1}\}, \ldots, \{\mathbf{F}_{sel,q}, \mathbf{S}_{sel,q}, \kappa_{sel,q}\}\}$, where $q = |\mathbf{F}_{prunable}|$, and the superscript $(l)$ denotes the number of the input models $\kappa_l$, taken from $\kappa$. Up to this point, the available configuration space has a size of $|\kappa| \cdot |\mathbf{F}_{prunable}|$. For each configuration resulting from Algorithm 3, the available precision configurations **nb** are swept, for a final space described by $\zeta = \langle \mathcal{F}, \mathcal{S}, \mathcal{M}, \mathcal{N} \rangle$ of size $|\kappa| \cdot |\mathbf{F}_{prunable}| \cdot |\mathcal{N}|$, where $\mathcal{N}$ contains the selected precision. The experimental section shows a workaround to reduce search space size and the steps required by the Pareto-optimal search. Regarding complexity, the feature selection in Algorithm 3 is a greedy search; thus, its complexity is linear in the number of features times the number of iterations needed for convergence to the desired accuracy or cost.[3] The PC pruning algorithm consists of an upward pass on the PC and its complexity is therefore linear in the size of the PC.

### 5.3.2　Pareto-Optimal Configuration Selection

Algorithm 2 is once again used to extract the Pareto-optimal configuration subset. The input is the configuration set $\zeta = \langle \mathcal{F}, \mathcal{S}, \mathcal{M}, \mathcal{N} \rangle$ and their corresponding

---

[3] In Algorithm 3 the user can provide the desired accuracy or cost as the while-loop break criterion.

accuracy (**acc**) and cost (**cost**). The output is the set of Pareto-optimal system configurations $\zeta^* = \{\{\kappa_i^*, \mathbf{F}^*_i, \mathbf{S}^*_i, nb_i^*\}_{i=1:p}\}$, each guaranteed to achieve the largest reachable accuracy for any given cost or the lowest reachable cost for any given accuracy (**acc**\*, **cost**\*).

The experiments in Sect. 5.4 illustrate how the proposed methodology can reap the benefits of considering system-wide hardware scalability.

## 5.4  Experiments: Pareto-Optimal Trade-off

The proposed techniques are empirically evaluated on a relevant embedded sensing use case: the human activity recognition (HAR) benchmark [9]. Additionally, the method's general applicability is shown on a number of other publicly available datasets [10–14], two of them commonly used for density estimation benchmarks and the rest commonly used for classification (see Table 5.1).

**Inference costs** Inference costs $C_M$ are based on the energy benchmarks introduced in [15] and [7] and discussed in Sect. 3.4. Table 3.2 shows the relative costs of each term in $C_M$ and how they scale with precision $nb$. The baseline is 64 floating point bits because it is the standard IEEE representation in software environments. For the rest of the experiments, three other standard low-precision representations are considered: 32 bits (8 exponent and 24 mantissa), 16 bits (5 exponent and 11 mantissa), and 8 bits (4 exponent and 4 mantissa) [16].

**Dataset Pre-processing** Numerical features were discretized with the method in [17] and binarized with a one-hot encoding. They were then subjected to a 75%-train, 10%-validation, and 15%-test random split. The time-series information of the HAR benchmark was preserved by using the first 85% samples for training and validation and the rest for testing. For the density estimation datasets, the splits provided in [13] were used, and the last variable in the set was assumed to be the class variable. All datasets underwent a process of wrapper feature selection (evaluating the features' value on a TAN classifier) before going through the hardware-aware optimization process to avoid over-fitting on the baseline model and ensure it is a fair reference point. The number of effectively used features $|\mathbf{F}|$ per benchmark is shown in Table 5.1. For datasets with less than 30 features, the prunable set $\mathbf{F}_{prun}$ is assumed to contain all the available features. For the rest, the 20 features with the highest correlation to the class variable are included in the prunable set. Within the context of an application, the prunable set can be user defined. For instance, in a multi-sensor seizure detection application, medical experts might advise against pruning features extracted from an EEG monitor.

**Model Learning** The models were learned on the train and validation sets ($\mathcal{F}_{\text{train}}$, $\mathcal{F}_{\text{valid}}$) with the LEARNPSDD algorithm [3], using the same settings reported therein and the following bottom-up vtree induction. To populate the model set $\kappa$, a model was retained after every $N_{\text{conv}}/10$ iterations, where $N_{\text{conv}}$ is the number of

**Table 5.1** Experimental
datasets

| Dataset | $|\mathbf{F}|$ | $|\mathbf{F}_{prun}|$ | $|\kappa|$ |
|---------|-----|---------|-----|
| 1. Banknote[a] | 15 | 15 | 11 |
| 2. HAR[a] | 28 | 28 | 11 |
| 3. Houses[a] | 36 | 20 | 11 |
| 4. Jester[b] | 99 | 20 | 11 |
| 5. Madelon[a] | 20 | 20 | 11 |
| 6. NLTCS[b] | 15 | 15 | 11 |
| 7. Six-HAR[a] | 54 | 20 | 11 |
| 8. Wilt[a] | 11 | 11 | 11 |

[a]Classification
[b] Density estimation

iterations needed for convergence (this is until the log-likelihood on validation data stagnates). Table 5.1 shows $|\kappa|$ for each dataset. Furthermore, as a baseline, a TAN classifier was trained and compiled to an arithmetic circuit.

### 5.4.1  Embedded Human Activity Recognition

The HAR dataset aims to recognize the activity a person is performing based on the statistical and energy features extracted from smartphone accelerometer and gyroscope data. The experiments herewith focus on a binary classification that discerns "walking downstairs" from the other activities. A total of 28 binary features were used, 8 of which are extracted from the gyroscope's signal and the rest from the accelerometer [18]. All inference costs for this dataset were normalized according to the energy consumption trends of an embedded ARM M9 CPU, assuming 0.1 nJ per operation [19].

**Sensor Costs** These costs were defined relative to inference costs. Specifically, each sensor is assumed to incur a cost equivalent to 10% of computational cost required to classify a single instance, as explained in the following sentences. Feature extraction and classification are assumed to take place in the aforementioned CPU that consumes, on average, 1 W to execute 10G operations per second. It will thus consume approximately 0.1 nJ per operation of the CPU.[4] Considering that the largest PC requires about 20,000 operations per instance classified, a total inference energy consumption of 2 μJ is assumed. Furthermore, both the gyroscope and the accelerometer are assumed to consume at least 2 mW when operating at 10 KHz, and they thus consume roughly 0.2 μJ per operation of the CPU.[5]

---

[4]https://developer.arm.com/products/processors/classic-processors.

[5]Gyroscopes can consume as much as 10 times more energy than accelerometers, but they are assumed to belong to a larger system, such as an inertial measurement unit. The same cost is therefore assigned to both.

**Feature Costs**  The features of this dataset are extracted by sampling the sensory signal, applying three low-pass filters, and calculating statistical quantities (mean, maximum/minimum, correlation, and standard deviation) on the resulting signal. Sampling and extracting the statistical features require a small number of operations in comparison to filtering. For example, calculating the mean of a sample requires a single MAC (multiply–accumulate, consisting of a multiplication and an addition) operation, whereas a third order low-pass filter will require at least nine. Filtering thus takes the bulk of the computations, so it is assumed that each feature extraction incurs a cost of 30 MAC operations.

**Pareto-Optimal Configuration**

This experiment consisted of three stages as shown in Fig. 5.3a for the training set $\mathcal{F}_{train}$: (1) Each model in $\kappa$ is first mapped to the trade-off space, as shown in black. (2) Starting from each model, the feature and sensor sets $(\mathbf{F}, \mathbf{S})$ were scaled with SCALESI, defined in Algorithm 3. The local Pareto-optimal extracted after this stage is shown in blue. (3) The precision $nb$ of each of these Pareto pruned configurations was then reduced, resulting in the Pareto front highlighted in red. As shown by configurations highlighted in green, the proposed method preserves the highest baseline train-set accuracy by pruning 11 of the available 28 features, which results in $C_{HA}$ savings of 53% (Point 2). When willing to tolerate further accuracy losses of 0.4%, the method outputs a configuration that consumes only 13% of the original cost by using a smaller model ($\kappa_3$), pruning 18 features, turning one sensor off and using a 32 bit representation (Point 3).

Figure 5.3b shows how the resulting configurations map to the test set, in contrast with a TAN classifier (blue star). The proposed methodology achieves the same accuracy as the TAN baseline but is capable of operating at a wide range of costs, with a variety of system configurations available.

Figure 5.4a and b breaks down the inference cost $C_M$ and the sensor cost $C_{SI}$. When considering only the costs of the PC evaluation (graph a), the proposed method results in savings of almost two orders of magnitude with accuracy losses lower than 1% and up to 3 orders of magnitude when tolerating more losses. Indeed, these inference cost savings result from the joint effects of precision reduction and from the effects of simplifying the PC with Algorithm 4. The histogram in Fig. 5.4c shows that, on average, every pruned feature procures $C_M$ savings of between 1% and 5% and up to 13%.

The sensor and feature costs shown in Fig. 5.4b only scale up to 50%, since at least one of the sensors must always be operating. This demonstrates the importance of taking these costs into account: even with significant inference costs savings, the system is still limited by the sensing costs.

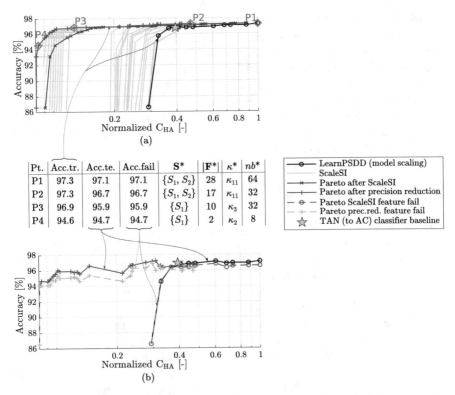

| Pt. | Acc.tr. | Acc.te. | Acc.fail | S* | $|\mathbf{F}^*|$ | $\kappa^*$ | $nb^*$ |
|-----|---------|---------|----------|------|--------|------------|--------|
| P1 | 97.3 | 97.1 | 97.1 | $\{S_1, S_2\}$ | 28 | $\kappa_{11}$ | 64 |
| P2 | 97.3 | 96.7 | 96.7 | $\{S_1, S_2\}$ | 17 | $\kappa_{11}$ | 32 |
| P3 | 96.9 | 95.9 | 95.9 | $\{S_1\}$ | 10 | $\kappa_3$ | 32 |
| P4 | 94.6 | 94.7 | 94.7 | $\{S_1\}$ | 2 | $\kappa_2$ | 8 |

Legend:
- LearnPSDD (model scaling)
- ScaleSI
- Pareto after ScaleSI
- Pareto after precision reduction
- Pareto ScaleSI feature fail
- Pareto prec.red. feature fail
- ☆ TAN (to AC) classifier baseline

**Fig. 5.3** Experiments on the human activity recognition benchmark. (**a**) Training set results, showing the trade-off from sequential scaling of system properties. (**b**) Test set results and results with unobserved features

## Robustness Against Failing Features

The proposed method was also evaluated in terms of its robustness to missing features, as shown in green and magenta in Fig. 5.3b. This was assessed by simulating, per Pareto-optimal model, ten iterations of random failure of varying sizes of feature sets ($|\mathbf{F}|/10, |\mathbf{F}|/5, |\mathbf{F}|/2$). The green and magenta dotted curves show the median of these experiments for the Pareto configurations and for the original model set. These trials stay within a range of $-2\%$ compared to the fully functional results in red and black, which validates the choice of a generative PSDD learner that can naturally cope with missing features at prediction time.

Section 5.5 proposes a learning strategy that forces feature variables to be conditioned on the class variable, inspired by the structure of Bayesian network classifiers. This learning strategy continues to reap the benefits that generative learning has on robustness to missing features but ensures that the conditional distribution $\Pr(\mathbf{F}|C)$ relevant to the classification task is encoded by the learned

**Fig. 5.4** (**a**)Accuracy versus inference costs $C_M$. (**b**) Accuracy versus sensor interfacing costs $C_{SI}$. (**c**) Inference cost $C_M$ savings per feature pruned

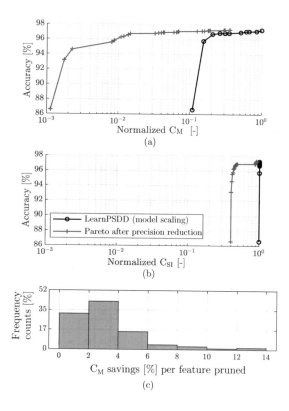

model. This can deliver models performing at higher accuracy than TAN classifiers while remaining robust to missing features.

## 5.4.2   Generality of the Method: Evaluation on Benchmark Datasets

The general applicability of the proposed strategy was evaluated on the remaining datasets, specified in Table 5.1. Due to the lack of information on the hardware that originated these datasets, the experiments in this section only consider inference costs $C_M$, again evaluated on the cost model from Sect. 3.4 with the values specified at the beginning of this section. Table 5.2 shows this cost along with the training and testing accuracies ($Acc_{tr}$, $Acc_{te}$) at four operating points for every dataset. Note that the six-class HAR benchmark has also been included here, to demonstrate the applicability of the proposed method beyond binary classification.

Figures 5.5 and 5.6 show the Pareto fronts for all the datasets in Table 5.1. The black line represents the models complexity scaling step, the blue one represents the sensor interface scaling stage, and the red one the final Pareto-optimal config-

**Fig. 5.7** Vtree for an $n$-value classification problem. (**a**) Variable $C$. (**b**) Propositional variables for each value of $C$

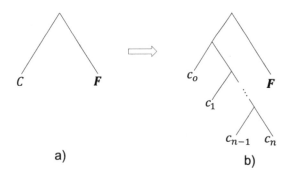

a)                                                                          b)

$$[root] = \bigvee_{i=0...n} ([c_i \bigwedge_{j:0...n \wedge i \neq j} \neg c_j] \wedge [s_i]). \tag{5.8}$$

In the case of a binary classification task this can be simplified to $[root] = ([\neg c] \wedge [s_0]) \vee ([c] \wedge [s_1])$. The variable decomposition described above implies that the prime of the vtree's root node is a right linear binary tree over the propositional variable $c_i$ and that the feature set $\mathbf{F}$ is indeed the sub variable as shown in Fig. 5.7.

Going back to the goal distribution in Eq. (5.7), it is necessary that the term $\Pr_{s_i}(\mathbf{F})$ represents the conditional relation of interest between $\mathbf{F}$ and $C$. This is achieved with the following proposition:

**Proposition 5.1** *Given (i) a vtree with variable $C$ as the prime and variables $\mathbf{F}$ as the sub of the root node, and (ii) an initial PSDD where the root decision node decomposes the distribution as $[root] = \bigvee_{i:0...n}([p_i] \wedge [s_i])$, applying the split and clone operators will never change the root decision decomposition $[root] = \bigvee_{i:0...n}([p_i] \wedge [s_i])$.*

**Proof** The D-LEARNPSDD algorithm iteratively applies two operations: split and clone (following the LEARNPSDD algorithm in [3]). First, the clone operation requires a parent node that is not available for the root node. The nodes corresponding to the prime of the root node $c_0, \ldots, c_n$ cannot be cloned either because the structure of the initial PSDD would not support the necessary copy and rerouting actions. Second, the split operator splits AND nodes into multiple elements by constraining the prime with a partial assignment to the prime variable. These partial assignments are mutually exclusive and exhaustive to keep the decision node deterministic. In the present case, the logical formula at the root of the PSDD already encodes mutual exclusivity among the possible assignments of the class (prime) variable. The root decision node is therefore already split into all the possible worlds available and the split operator would be inconsequential. For these reasons, the root's base remains intact.                                                        □

The following proposition and examples show that the resulting PSDD includes nodes that directly represent the distribution $\Pr(\mathbf{F}|C)$.

**Proposition 5.2** *A PSDD of the form* $\bigvee_{i=0...n}([c_i \bigwedge_{j:0...n \wedge i \neq j} \neg c_j] \wedge [s_i])$ *with* $s_i$ *any formula with propositional feature variables* $f_0, \ldots, f_n$ *directly expresses the distribution* $\Pr(\mathbf{F}|C)$.

***Proof*** Applying this to Eq. (5.1) results in

$$\Pr_{root}(C\mathbf{F}) = \sum_i \theta_i \Pr_{c_i}(C) \Pr_{s_i}(\mathbf{F})$$

$$= \sum_i \theta_i \Pr_{c_i}\left(C|[c_i \bigwedge_{j:0...n \wedge j \neq i} \neg c_j]\right) \cdot \Pr_{s_i}\left(\mathbf{F}|[c_i \bigwedge_{j:0...n \wedge j \neq i} \neg c_j]\right)$$

$$= \sum_i \theta_i \Pr_{c_i}(C = i) \cdot \Pr_{s_i}(\mathbf{F}|C = i).$$

The learned PSDD comprises $n$ nodes $s_i$ with distribution $\Pr_{s_i}$ that directly represents $\Pr(\mathbf{F}|C = i)$. The PSDD thus encodes $\Pr(\mathbf{F}|C)$ directly because there are $n$ possible value assignments of $C$.                                                       □

In addition to defining a suitable vtree, it is necessary to define an initial PSDD that is consistent with the logic formulas above, as illustrated by the following examples. These examples consider binary classification, but the concepts extend to the $n$-class scenario.

*Example 5.1* Figure 5.8b shows a PSDD that encodes a fully factorized probability distribution normalized for the vtree in Fig. 5.8a. Note that the vtree does not connect the class variable $C$ to all feature variables (e.g. $F_1$) so D-LEARNPSDD is not guaranteed to find conditional relations between certain features and the class when initialized with this PSDD.

*Example 5.2* Figure 5.8e shows a PSDD that explicitly conditions the feature variables on the class variables by normalizing for the vtree in Fig. 5.8c and by following the logical formula from Proposition 5.2.

*Example 5.3* When initializing on a PSDD that encodes a fully factorized formula (Fig. 5.8d), the base of the root node is $[root] = [c \vee \neg c] \wedge [s_0]$. Equation (5.1) evaluates to

$$\Pr_q(C\mathbf{F}) = \Pr_{p_0}(C|[c \vee \neg c]) \cdot \Pr_{s_0}(\mathbf{F}|[c \vee \neg c])$$

$$= \left(\Pr_{p_1}(C|[c]) + \Pr_{p_2}(C|[\neg c])\right) \cdot \Pr_{s_0}(\mathbf{F}|[c \vee \neg c])$$

$$= \left(\Pr_{p_1}(C = 1) + \Pr_{p_2}(C = 0)\right) \cdot \Pr_{s_0}(\mathbf{F}).$$

Thus, in this worst case scenario, the PSDD encodes a distribution that assumes the class to be independent from all feature variables, which may still lead to high log-likelihood but will have low classification accuracy.

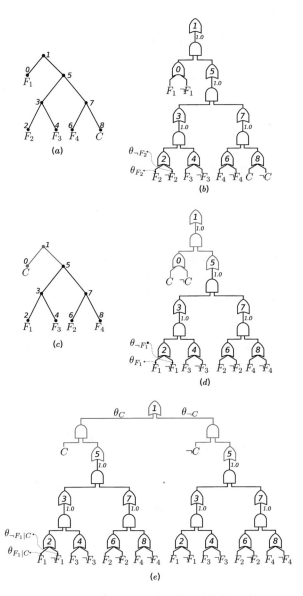

**Fig. 5.8** (a) Vtree learned by minimizing pairwise Mutual Information among variables. (b) PSDD corresponding to a fully factorized distribution, guided by the vtree in (a). (c) Class-variable constrained vtree, learned by minimizing Conditional Mutual Information. (d) PSDD corresponding to a fully factorized distribution, guided by the vtree in (c). (e) Class-constrained PSDD guided by the vtree in (c). This PSDD explicitly conditions feature variables on the class variable

## 5.5.2   Generative Bias and Vtree Learning

Learning the distribution over the feature variables is a generative learning process by applying the split and clone operators in the same way as the original LEARNPSDD algorithm. To define the nodes corresponding to $s_i$ with distributions $\Pr(\mathbf{F}|C = i)$ the intuition behind (TA)NB is followed: learning is initialized with a PSDD that encodes a distribution where all feature variables are independent given the class variable (see Fig. 5.8e). Next, the LEARNPSDD algorithm will incrementally learn the relations between the feature variables by applying the split and clone operations following the approach in [3].

In LEARNPSDD, the decision nodes decompose the distribution into independent distributions. Thus, the vtree is learned from data by minimizing the approximate pairwise mutual information, as this metric quantifies the level of independence between two sets of variables. D-LEARNPSDD is interested in the level of conditional independence between sets of feature variables given the class variable. The vtree is therefore obtained by optimizing for conditional mutual information ($CMI$) instead, which replaces mutual information in the approach in [3] with

$$CMI(\mathbf{X}, \mathbf{Y}|\mathbf{Z}) = \sum_{\mathbf{x}} \sum_{\mathbf{y}} \sum_{\mathbf{z}} \Pr(\mathbf{xy}) \log \frac{\Pr(\mathbf{z}) \Pr(\mathbf{xyz})}{\Pr(\mathbf{xz}) \Pr(\mathbf{yz})}. \tag{5.9}$$

## 5.6   Experiments: Biased PSDD Learning

This section compares the performance of D-LEARNPSDD, LEARNPSDD, two generative Bayesian classifiers (NB and TANB), and a discriminative classifier (logistic regression): (1) Section 5.6.2 examines whether the introduced discriminative bias improves classification performance on PSDDs. (2) Section 5.6.3 analyzes the impact of the vtree and the imposed structural constraints on model tractability and performance. (3) Finally, Sect. 5.6.4 compares the robustness to missing values for all classification approaches.

### 5.6.1   Experimental Setup

Experiments are performed on the suite of 15 standard machine learning benchmarks listed in Table 5.3, which also summarizes the number of binary features $|\mathbf{F}|$, the number of classes $|C|$, and the available number of training samples $|N|$ per dataset. All of the datasets come from the UCI machine learning repository [10], with exception of "Mofn" and "Corral" [22]. The datasets are pre-processed by applying the discretization method described in [17]. Variables are then binarized

**Table 5.3** Dataset details for robust classification with PSDDs

| Dataset | **\|F\|** | **\|C\|** | **\|N\|** |
|---|---|---|---|
| 1. Australian | 40 | 2 | 690 |
| 2. Breast | 28 | 2 | 683 |
| 3. Chess | 39 | 2 | 3196 |
| 4. Cleve | 25 | 2 | 303 |
| 5. Corral | 6 | 2 | 160 |
| 6. Credit | 42 | 2 | 653 |
| 7. Diabetes | 11 | 2 | 768 |
| 8. German | 54 | 2 | 1000 |
| 9. Glass | 17 | 6 | 214 |
| 10. Heart | 9 | 2 | 270 |
| 11. Iris | 12 | 3 | 150 |
| 12. Mofn | 10 | 2 | 1324 |
| 13. Pima | 11 | 2 | 768 |
| 14. Vehicle | 57 | 2 | 846 |
| 15. Waveform | 109 | 3 | 5000 |

using a one-hot encoding. Moreover, instances with missing values and features whose value was always equal to 0 were removed.

## 5.6.2  Evaluation of D-LearnPSDD

Table 5.4 compares D-LEARNPSDD, LEARNPSDD, naive Bayes (NB), Tree Augmented Naive Bayes (TANB), and logistic regression (LogReg)[6] in terms of accuracy via five-fold cross-validation[7] and in terms of size with the number of parameters. For LEARNPSDD, a model is incrementally learned on each fold until convergence on validation data log-likelihood, following the methodology in [3].

For D-LEARNPSDD, a model was learned incrementally on each fold until likelihood converged. The model with the highest training set accuracy was then selected for the experiment. For NB and TANB, a model was learned per fold and compiled to arithmetic circuit representations,[8] a more general form of PSDDs [23], which allow to compare the sizes of Bayes Net classifiers and the PSDDs. Finally, all probabilistic models are compared to a discriminative classifier: a multinomial logistic regression model with a ridge estimator.

Table 5.4 highlights, in bold, the best performing method for each dataset. It is clear that the proposed D-LEARNPSDD outperforms LEARNPSDD in all but two datasets, as the latter method is not guaranteed to learn significant relations

---

[6]NB, TANB, and LogReg are learned using Weka with default settings.

[7]In each fold, 10% of the data is held for validation.

[8]Using the ACE tool Available at http://reasoning.cs.ucla.edu/ace/.

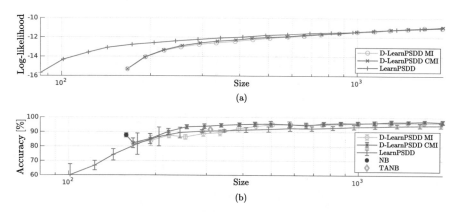

**Fig. 5.9** Log-likelihood (**a**) and accuracy (**b**) vs. model size trade-off of the incremental PSDD learning approaches. MI and CMI denote mutual information and conditional mutual information vtree learning, respectively

between feature and class variables. Moreover, it outperforms Bayesian classifiers in most benchmarks, as the learned PSDDs are more expressive and allow to encode complex relationships among sets of variables or local dependencies such as context specific independence while remaining tractable. Finally, note that the D-LEARNPSDD is competitive in terms of accuracy with respect to logistic regression (LogReg), a purely discriminative classification approach.

### 5.6.3   Impact of the Vtree on Discriminative Performance

The structure and size of the learned PSDD are largely determined by the vtree it is normalized for. Naturally, the vtree also has an important role in determining the quality (in terms of log-likelihood) of the probability distribution encoded by the learned PSDD  [3]. This section studies the impact that the choice of vtree and learning strategy has on the trade-offs between model size, quality, and discriminative performance.

Figure 5.9a shows test set log-likelihood and Fig. 5.9b classification accuracy as a function of model size (in number of parameters) for the "Chess" dataset. Average log-likelihood and accuracy are displayed over logarithmically distributed ranges of model size. This figure contrasts the results of three learning approaches: D-LEARNPSDD when the vtree learning stage optimizes mutual information (MI, shown in light blue); when it optimizes conditional mutual information (CMI, shown in dark blue); and the traditional LEARNPSDD (in orange).

Figure 5.9a shows that likelihood improves at a faster rate during the first iterations of LEARNPSDD but eventually settles to the same values as D-LEARNPSDD because both optimize for log-likelihood. However, the discriminative bias guarantees that classification accuracy on the initial model will be at least as high

**Table 5.4** Five cross-fold accuracy and size in number of parameters for the datasets in Table 5.3

| Dataset | D-LearnPSDD Accuracy | D-LearnPSDD Size | LearnPSDD Accuracy | LearnPSDD Size | NB Accuracy | NB Size | TANB Accuracy | TANB Size | LogReg Size | LogReg Accuracy |
|---|---|---|---|---|---|---|---|---|---|---|
| 1. Australian | **86.2 ± 3.6** | 367 | 84.9 ± 2.7 | 386 | 85.1 ± 3.1 | 161 | 85.8 ± 3.4 | 312 | 312 | 84.1 ± 3.4 |
| 2. Breast | 97.1 ± 0.9 | 291 | 94.9 ± 0.5 | 491 | **97.7 ± 1.2** | 114 | 97.7 ± 1.2 | 219 | 219 | 96.5 ± 1.6 |
| 3. Chess | **97.3 ± 1.4** | 2178 | 94.9 ± 1.6 | 2186 | 87.7 ± 1.4 | 158 | 91.7 ± 2.2 | 309 | 309 | 96.9 ± 0.7 |
| 4. Cleve | 82.2 ± 2.5 | 292 | 81.9 ± 3.2 | 184 | **84.9 ± 3.3** | 102 | 79.9 ± 2.2 | 196 | 196 | 81.5 ± 2.9 |
| 5. Corral | **99.4 ± 1.4** | 39 | 98.1 ± 2.8 | 58 | 89.4 ± 5.2 | 26 | 98.8 ± 1.7 | 45 | 45 | 86.3 ± 6.7 |
| 6. Credit | 85.6 ± 3.1 | 693 | 86.1 ± 3.6 | 611 | **86.8 ± 4.4** | 170 | 86.1 ± 3.9 | 326 | 326 | 84.7 ± 4.9 |
| 7. Diabetes | **78.7 ± 2.9** | 124 | 77.2 ± 3.3 | 144 | 77.4 ± 2.56 | 46 | 75.8 ± 3.5 | 86 | 86 | 78.4 ± 2.6 |
| 8. German | 72.3 ± 3.2 | 1185 | 69.9 ± 2.3 | 645 | 73.5 ± 2.7 | 218 | **74.5 ± 1.9** | 429 | 429 | 74.4 ± 2.3 |
| 9. Glass | **79.1 ± 1.9** | 214 | 72.4 ± 6.2 | 321 | 70.0 ± 4.9 | 203 | 69.5 ± 5.2 | 318 | 318 | 73.0 ± 5.7 |
| 10. Heart | **84.1 ± 4.3** | 51 | 78.5 ± 5.3 | 75 | 84.0 ± 3.8 | 38 | 83.0 ± 5.1 | 70 | 70 | 84.0 ± 4.7 |
| 11. Iris | 90.0 ± 0.1 | 76 | 94.0 ± 3.7 | 158 | **94.7 ± 1.8** | 75 | 94.7 ± 1.8 | 131 | 131 | 94.7 ± 2.9 |
| 12. Mofn | 98.9 ± 0.9 | 260 | 97.1 ± 2.4 | 260 | 85.0 ± 5.7 | 42 | 92.8 ± 2.6 | 78 | 78 | **100.0 ± 0** |
| 13. Pima | **80.2 ± 0.3** | 108 | 74.7 ± 3.2 | 110 | 77.6 ± 3.0 | 46 | 76.3 ± 2.9 | 86 | 86 | 77.7 ± 2.9 |
| 14. Vehicle | **95.0 ± 1.7** | 1186 | 93.9 ± 1.69 | 1560 | 86.3 ± 2.00 | 228 | 93.0 ± 0.8 | 442 | 442 | 94.5 ± 2.4 |
| 15. Waveform | 85.0 ± 1.0 | 3441 | 78.7 ± 5.6 | 2585 | 80.7 ± 1.9 | 657 | 83.1 ± 1.1 | 1296 | 1296 | **85.5 ± 0.7** |

Bold values denote the best performing method per dataset

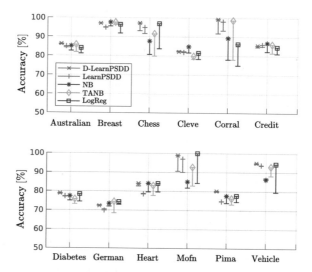

**Fig. 5.10** Classification robustness against missing features per method

as that of a Naive Bayes classifier (see Fig. 5.9b). Moreover, this results in consistently superior accuracy (for the CMI case) compared to the purely generative LEARNPSDD approach as shown also in Table 5.4. The dip in accuracy during the second and third intervals is a consequence of the generative learning that optimizes for log-likelihood and can therefore initially yield feature-value correlations that decrease the model's performance as a classifier.

Finally, Fig. 5.9b demonstrates that optimizing the vtree for conditional mutual information results in overall better performance vs. accuracy trade-off when compared to optimizing for mutual information. Such a conditional mutual information objective function is consistent with the conditional independence constraint imposed on the structure of the PSDD and allows the model to consider the special status of the class variable in the discriminative task.

### 5.6.4   Robustness to Missing Features

Generative models encode a joint probability distribution over all variables and therefore tend to be more robust against missing features than discriminative models that only learn relations relevant to their discriminative task. This experiment assesses this robustness aspect by simulating the random failure of 10% of the original feature set per benchmark and per fold in five-fold cross-validation. Figure 5.10 shows the average accuracy over 10 such feature failure trials in each of the five folds (flat markers) in relation to their full feature set accuracy reported in Table 5.4 (shaped markers). As expected, the performance of the discriminative classifier (LogReg) suffers the most during feature failure, while D-LEARNPSDD and LEARNPSDD are notably more robust than any other approach, with accuracy losses of no more than 8%. Note from the flat markers that the performance of D-LEARNPSDD under feature failure is the best in all datasets but one.

The robustness and improved discriminative performance brought about by D-LEARNPSDD will be exploited by the run-time strategy presented in Chap. 6 that tunes the complexity of the used model dynamically.

## 5.7  Related Work

Tractable learning aims to balance the trade-off between how well the resulting models fit the available data and how efficiently queries are answered. Most implementations focus on maximizing model performance and only factor in query efficiency by subjecting the learning stage to a fixed tractability constraint (e.g. max treewidth [24]). While recent notions of tractability consider the cost of probabilistic inference as the number of arithmetic operations involved in a query [25, 26], they still disregard hardware implementation nuances of the target application.

There are, however, many examples of hardware-efficient probabilistic inference realizations, addressed by both the probabilistic models and the embedded hardware communities from several perspectives. The works by Tschiatschek and Pernkopf [27] and Piatkowski et al. [28] propose reduced precision and integer representation schemes for PGMs as a strategy to address the constraints of embedded devices. Other efficient hardware implementation efforts have been made by Khan and Wentzloff [29], where a hardware accelerator that targets inference in PGMs is proposed, while Zermani et al. [30], Schumann et al. [31], and Sommer et al. [32] have proposed to accelerate inference on SPNs, capitalizing on their tractable properties.

The strategies proposed throughout this chapter constitute an effort to integrate the expertise from both communities under a unified framework that considers the impact of all scalable aspects of the model to optimize it in a hardware-aware fashion. To that end, it leverages the properties of the selected PC representation. Such representation enables the use of the proposed framework with any probabilistic model that is compute-efficient at prediction time: see [33] by Holtzen et al. and [34] by Vlasselaer et al. for examples of probabilistic program compilation to ACs and [26] by Lowd and Rooshenas on how to perform efficient inference with Markov networks represented as ACs.

Regarding the conflict between generative and discriminative model learning, that the second part of this chapter addresses, there is a vast body of relevant literature, some dating back decades [35]. There are multiple techniques that support learning of parameters [36, 37] and structure [38, 39] of Probabilistic Circuits. Typically, different approaches are followed to learn either generative or discriminative tasks, but some methods exploit discriminative models' properties to deal with missing variables [40].

Other works that also constrain the structure of PSDDs have been proposed before, such as Choi et al. [41]. However, they only do parameter learning, not structure learning: their approach to improve accuracy is to learn separate structured PSDDs for each distribution of features given the class and feed them to a NB classifier. In [42], Correia and De Campos propose a constrained SPN architecture that shows both computational efficiency and classification performance improvements. However, it focuses on decision robustness rather than robustness against missing values, essential to the application range discussed in this chapter.

There are also a number of methods that focus specifically on the interaction between discriminative and generative learning. In [43], Khosravi et al. provide a method to compute expected predictions of a discriminative model with respect to a probability distribution defined by an arbitrary generative model in a tractable manner. This combination allows to handle missing values using discriminative counterparts of generative classifiers [44]. More distant to the work proposed herewith is the line of hybrid discriminative and generative models [45], and their focus is on semi-supervised learning and deals with missing labels.

## 5.8  Discussion

This chapter has discussed works that constitute one of the first efforts to introduce the field of tractable probabilistic reasoning to the emerging domain of edge computing. Sections 5.2–5.4 discussed the proposal of a novel hardware-aware Probabilistic Circuit that deals with the limitations of the efficiency versus performance trade-off considered by the field of tractable learning. The proposed method relies on the notion of hardware-aware cost proposed in Chap. 3 and obtains the Pareto-optimal system-configuration set in the hardware-aware cost versus accuracy space by means of a sequential hardware-aware search and a Pareto-optimal configuration selection stage. Experiments on a variety of benchmarks demonstrated the effectiveness of the approach and sacrifice little to no accuracy for significant cost savings. This opens up opportunities for the efficient implementation of probabilistic models in resource-constrained edge devices.

The second part of this chapter (Sect. 5.5) introduced a PSDD learning technique that improves classification performance by introducing a discriminative bias. Meanwhile, robustness against missing data is kept by exploiting generative learning. The method capitalizes on PSDDs' domain knowledge encoding capabilities to enforce the conditional relation between the class and the features. It is proven that this constraint is guaranteed to be enforced throughout the learning process and it is shown how not encoding such a relation might lead to poor classification performance. Evaluation on a suite of benchmarking datasets in Sect. 5.6 shows that the proposed technique outperforms purely generative PSDDs in terms of classification accuracy and the other baseline classifiers in terms of robustness.

The two lines of research above open up many opportunities for hardware-aware realizations under dynamic real-world scenarios. The hardware-aware cost metric enables the algorithmic level of abstraction to access other system levels and therefore exploits scalability opportunities that might otherwise not be available. This is of particular relevance to dynamic scenarios where the scalability aspect can be leveraged to, for example, extend battery life or make a smarter use of the available resources while meeting user defined performance needs. These needs can, in turn, be met by an approach that considers the discriminative task at hand, just like the proposed hybrid learning strategy proved to be capable of. But the scalability aspect can only be ensured if the algorithm remains robust to missing features and noise. The following chapter addresses some of these aspects by means of dynamic model complexity scaling strategies.

As future work, besides the use case of allowing Probabilistic Circuits to run on resource-constrained devices, one could consider the cost not only in terms of energy savings but also time. Considering, for example, applications like self-driving cars, one could do an early prediction of critical conditions even before having the measurements from all the sensors. In these types of applications, latency is a critical resource to take into consideration, as was discussed at the beginning of this book in Sect. 1.1.1.

# References

1. A. Darwiche, P. Marquis, A knowledge compilation map. J. Artif. Intell. Res. **17**, 229–264 (2002)
2. A. Darwiche, SDD: a new canonical representation of propositional knowledge bases, in *International Joint Conference on Artificial Intelligence*, 2011
3. Y. Liang, J. Bekker, G. Van den Broeck, Learning the structure of probabilistic sentential decision diagrams, in *Proceedings of the Conference on Uncertainty in Artificial Intelligence (UAI)*, 2017
4. D. Kisa, G. Van den Broeck, A. Choi, A. Darwiche, Probabilistic sentential decision diagrams, in *Fourteenth International Conference on the Principles of Knowledge Representation and Reasoning*, 2014
5. C. Boutilier, N. Friedman, M. Goldszmidt, D. Koller, Context-specific independence in Bayesian networks, in *Proceedings of the International Conference on Uncertainty in Artificial Intelligence*, 1996
6. J. Bekker, J. Davis, A. Choi, A. Darwiche, G. Van den Broeck, Tractable learning for complex probability queries, in *Advances in Neural Information Processing Systems*, 2015
7. N. Shah, L.I. Galindez Olascoaga, W. Meert, M. Verhelst, ProbLP: a framework for low-precision probabilistic inference, in *Proceedings of the 56th Annual Design Automation Conference 2019*, 2019, pp. 1–6
8. D. Buchfuhrer, C. Umans, The complexity of Boolean formula minimization, in *International Colloquium on Automata, Languages, and Programming* (Springer, New York, 2008), pp. 24–35
9. D. Anguita, A. Ghio, L. Oneto, X. Parra, J.L. Reyes-Ortiz, A public domain dataset for human activity recognition using smartphones, in *ESANN*, 2013
10. D. Dua, C. Graff, UCI machine learning repository, 2017 [Online]. Available: http://archive.ics.uci.edu/ml
11. I. Guyon, S. Gunn, NIPS feature selection challenge, 2003

12. B.A. Johnson, R. Tateishi, N.T. Hoan, A hybrid pansharpening approach and multiscale object-based image analysis for mapping diseased pine and oak trees. Int. J. Remote Sens. **34**(20), 6969–6982 (2013)
13. D. Lowd, J. Davis, Learning Markov network structure with decision trees, in *2010 IEEE International Conference on Data Mining* (IEEE, New York, 2010), pp. 334–343
14. R.K. Pace, R. Barry, Sparse spatial autoregressions. Stat. Probabil. Lett. **33**(3), 291–297 (1997)
15. M. Horowitz, 1.1 computing's energy problem (and what we can do about it), in *2014 IEEE International Solid-State Circuits Conference Digest of Technical Papers (ISSCC)*, February 2014, pp. 10–14
16. IEEE, IEEE standard for floating-point arithmetic. IEEE Std 754-2008, August 2008, pp. 1–70
17. U. Fayyad, K. Irani, Multi-interval discretization of continuous-valued attributes for classification learning, in *IJCAI*, 1993
18. L.I. Galindez Olascoaga, W. Meert, N. Shah, M. Verhelst, G. Van den Broeck, Towards hardware-aware tractable learning of probabilistic models, in *Advances in Neural Information Processing Systems (NeurIPS)*, vol. 32, December 2019
19. S. Tarkoma, M. Siekkinen, E. Lagerspetz, Y. Xiao, *Smartphone Energy Consumption: Modeling and Optimization* (Cambridge University Press, Cambridge, 2014)
20. N. Friedman, D. Geiger, M. Goldszmidt, Bayesian network classifiers. J. Mach. Learn. **29**(2), 131–163 (1997)
21. L.I. Galindez Olascoaga, W. Meert, N. Shah, G. Van den Broeck, M. Verhelst, Discriminative bias for learning probabilistic sentential decision diagrams, in *Proceedings of the Symposium on Intelligent Data Analysis (IDA)*, April 2020
22. R. Kohavi, G.H. John, Wrappers for feature subset selection. Artif. Intell. **97**(1–2), 273–324 (1997)
23. A. Darwiche, *Modeling and Reasoning with Bayesian Networks* (Cambridge University Press, Cambridge, 2009)
24. F.R. Bach, M.I. Jordan, Thin junction trees, in *Advances in Neural Information Processing Systems*, 2002, pp. 569–576
25. D. Lowd, P. Domingos, Learning arithmetic circuits, in *Proceedings of the Conference on Uncertainty in Artificial Intelligence (UAI)*, 2008
26. D. Lowd, A. Rooshenas, Learning Markov networks with arithmetic circuits, in *Artificial Intelligence and Statistics*, 2013, pp. 406–414
27. S. Tschiatschek, F. Pernkopf, On Bayesian network classifiers with reduced precision parameters. IEEE Trans. Patt. Anal. Mach. Intell. **37**(4), 774–785 (2015)
28. N. Piatkowski, S. Lee, K. Morik, Integer undirected graphical models for resource-constrained systems. Neurocomputing **173**, 9–23 (2016)
29. O.U. Khan, D.D. Wentzloff, Hardware accelerator for probabilistic inference in 65-nm CMOS. IEEE Trans. Very Large Scale Integr. (VLSI) Syst. **24**(3), 837–845 (2016)
30. S. Zermani, C. Dezan, R. Euler, J.-P. Diguet, Bayesian network-based framework for the design of reconfigurable health management monitors, in *2015 NASA/ESA Conference on Adaptive Hardware and Systems (AHS)* (IEEE, New York, 2015), pp. 1–8
31. J. Schumann, K.Y. Rozier, T. Reinbacher, O.J. Mengshoel, T. Mbaya, C. Ippolito, Towards real-time, on-board, hardware-supported sensor and software health management for unmanned aerial systems. Int. J. Prognost. Health Manage. (2015). ISSN: 2153-2648
32. L. Sommer, J. Oppermann, A. Molina, C. Binnig, K. Kersting, A. Koch, Automatic mapping of the sum-product network inference problem to FPGA-based accelerators, in *2018 IEEE 36th International Conference on Computer Design (ICCD)* (IEEE, New York, 2018), pp. 350–357
33. S. Holtzen, T. Millstein, G. Van den Broeck, Symbolic exact inference for discrete probabilistic programs, in *Proceedings of the ICML Workshop on Tractable Probabilistic Modeling (TPM)*, June 2019
34. J. Vlasselaer, G. Van den Broeck, A. Kimmig, W. Meert, L. De Raedt, TP-compilation for inference in probabilistic logic programs. Int. J. Approx. Reason. **78**, 15–32 (2016)
35. T. Jaakkola, D. Haussler, Exploiting generative models in discriminative classifiers, in *Advances in Neural Information Processing Systems*, 1999

36. H. Poon, P. Domingos, Sum-product networks: a new deep architecture, in *2011 IEEE International Conference on Computer Vision Workshops (ICCV Workshops)* (IEEE, New York, 2011), pp. 689–690
37. R. Gens, P. Domingos, Discriminative learning of sum-product networks, in *Advances in Neural Information Processing Systems*, 2012
38. A. Rooshenas, D. Lowd, Discriminative structure learning of arithmetic circuits, in *Artificial Intelligence and Statistics*, 2016, pp. 1506–1514
39. Y. Liang, G. Van den Broeck, Learning logistic circuits, in *Proceedings of the 33rd Conference on Artificial Intelligence (AAAI)*, 2019
40. R. Peharz, A. Vergari, K. Stelzner, A. Molina, X. Shao, M. Trapp, K. Kersting, Z. Ghahramani, Random sum-product networks: a simple and effective approach to probabilistic deep learning, in *Proceedings of the Conference on Uncertainty in Artificial Intelligence (UAI)*, 2019
41. A. Choi, N. Tavabi, A. Darwiche, Structured features in naive Bayes classification, in *Thirtieth AAAI Conference on Artificial Intelligence*, 2016
42. A.H. Correia, C.P. de Campos, Towards scalable and robust sum-product networks, in *International Conference on Scalable Uncertainty Management* (Springer, New York, 2019), pp. 409–422
43. P. Khosravi, Y. Choi, Y. Liang, A. Vergari, G. Van den Broeck, On tractable computation of expected predictions, in *Advances in Neural Information Processing Systems (NeurIPS)*, 2019
44. P. Khosravi, Y. Liang, Y. Choi, G. Van den Broeck, What to expect of classifiers? Reasoning about logistic regression with missing features, in *Proceedings of the 28th International Joint Conference on Artificial Intelligence (IJCAI)*, August 2019 [Online]. Available: http://starai.cs.ucla.edu/papers/KhosraviIJCAI19.pdf
45. J.A. Lasserre, C.M. Bishop, T.P. Minka, Principled hybrids of generative and discriminative models, in *IEEE Computer Society Conference on Computer Vision and Pattern Recognition (CVPR)*, 2006

# Chapter 6
# Run-Time Strategies

Chapters 4 and 5 proposed hardware-aware probabilistic models that can represent scalable device properties at different stages of the sensing embedded pipeline. These models can infer the probability of a certain outcome or observation given a device configuration of interest, which can be specified in terms of the signal quality procured by the sensor front-end, the type and number of extracted features, the precision used for arithmetic operations, etc. As such, hardware-aware probabilistic models can aid in evaluating the impact that a variety of such device configurations may have on the quality of the classification task, as well as on the system's cost. In turn, the models can be leveraged to determine the device configuration that will lead to a desired classification accuracy versus cost operating point.

The strategies proposed so far have focused on off-line optimization, where the goal is identifying the system configuration that will lead to the desired cost versus accuracy performance when deployed at application time. These strategies assume that run-time conditions will not differ significantly from the conditions under which the selection was made, and that the expected cost and accuracy will be met. However, in many applications, run-time conditions can change dynamically, resulting in situations not accounted for by the off-line optimization. Think, for example, of a self-driving vehicle whose sensors suddenly become unreliable due to unfavorable environmental conditions.

This chapter considers some of the implications of run-time embedded sensing scenarios where environmental conditions and system requirements may change over time. Two main aspects are addressed by the strategies proposed in this Chapter: robustness to missing features from, for example, malfunctioning sensors or occlusive environmental conditions, and the problem of coping with significant resource constraints while attempting to meet the accuracy requirements of the application.

A common thread throughout this book has been that the properties of probabilistic models make them amenable to scenarios where input variables might be noisy or unobserved. Indeed, this robustness has been demonstrated empirically

© The Author(s), under exclusive license to Springer Nature Switzerland AG 2021   111
L. I. Galindez Olascoaga et al., *Hardware-Aware Probabilistic Machine Learning Models*,
https://doi.org/10.1007/978-3-030-74042-9_6

for PSDDs in Sect. 5.6.4 but has also been exploited throughout Chaps. 4 and 5 to select device configurations that lead to significant cost savings with minimal accuracy losses. This chapter continues not only to rely on these properties to select Pareto-optimal configurations but also to address some of the challenges of dynamic run-time conditions.

Section 6.1 shows how the trade-off search strategy discussed in Chap. 4 can be adapted to account for sensors failing at run-time. This approach is empirically verified on a human activity recognition benchmark in Sect. 6.1.2, which includes results previously published in:

Galindez, L., Badami, K., Vlasselaer, J., Meert, W., and Verhelst, M. (2018). Dynamic Sensor-Frontend Tuning for Resource Efficient Embedded Classification. *IEEE Journal on Emerging and Selected Topics in Circuits and Systems*, 8(4), 858–872.

Section 6.2 addresses some of the practical limitations of that approach and proposes a model complexity tuning strategy that identifies the difficulty of the classification task on a given instance and dynamically switches to a higher or lower complexity setting accordingly. This strategy utilizes models learned with the techniques discussed in Chap. 5. It is shown that a large range of cost versus accuracy operating points close to the off-line Pareto front can be reached by only switching between two models, reducing the need to store and deploy a large array of models at run-time. It is also demonstrated in Sect. 6.3.4 that this adaptive behavior can deal more robustly with missing sensors than deploying a single off-line selected configuration. Some aspects of this strategy were previously presented in:

Galindez Olascoaga, L. I., Meert, W., Shah, N., and Verhelst, M. (2020b). Dynamic Complexity Tuning for Hardware-Aware Probabilistic Circuits. *In IoT, Edge, and Mobile for Embedded Machine Learning (ITEM) Workshop, collocated with ECML-PKDD 2020.*

Finally, Sects. 6.4 and 6.5 discuss some of the related literature and offer conclusions and suggestions for future work.

## 6.1   Noise-Scalable Bayesian Networks: Missing Features at Run-Time

Noise-scalable Bayesian network classifiers (ns-BNs) are equipped to represent the probabilistic relations among the signal quality variations offered by the scalable sensor front-end. Chapter 4 described multiple use cases that leverage the properties of this model to comprehensively explore the cost versus accuracy trade-off space and find Pareto-optimal front-end noise-tolerance settings. However, dynamically changing device and environmental conditions that are not represented in the model, such as the possibility of sensor malfunction, cannot be accounted for during off-line derivation of this single static Pareto-optimal set.

Even though probabilistic models can remain robust to unavailable features, owing to their ability to encode joint distributions and marginalize unobserved variables, the actual run-time conditions might differ from the conditions under which the Pareto-optimal selection was made. In that situation, the Pareto-optimal noise-tolerance configuration that was selected off-line may not exhibit the same performance at run-time and may even cease to achieve Pareto-optimal performance.

A suitable approach to account for these real-time changes is to, at run-time, locally determine appropriate noise-tolerance settings according to the current state of the device, instead of defining these settings beforehand, at an off-line stage. Section 6.1.1 describes how the Pareto-optimal feature quality tuning technique of Sect. 4.2 can be adapted to follow this run-time selection approach.

## 6.1.1  Run-Time Pareto-Optimal Selection

Recall that the strategy proposed in Chap. 4 determines which sensor front-end configurations lead to Pareto-optimal accuracy versus cost. However, these configurations are only guaranteed to be Pareto-optimal if the conditions under which they were selected remain unchanged. For example, Fig. 4.5 showed that the mixed-signal front-end costs on the human activity recognition benchmark could be reduced by more than four orders of magnitude with accuracy losses lower than 10% by turning the Gyroscope-X off and allowing more noise in the rest of the features. Suppose that the Accelerometer-X in the previous example is suddenly unavailable. Then it is likely that accuracy losses will be higher than the expected 10%. It is clear that the off-line selection approach proposed in Chap. 4 cannot account for unexpected run-time changes.

The solution proposed in this section is to evaluate the aforementioned strategy at run-time, such that the current state of the device is taken into consideration when setting the configuration of choice. For example, if the strategy can identify that one sensor is currently missing, then it will select a feature combination that does not require the usage of the failed or deactivated sensor. This is shown in Fig. 6.1, where the dynamic tuning block receives as input the state of the sensory chains $(S_1, S_2)$, which constitutes the initial configuration $\theta_{init}$ of the SCALEFEATURENOISE strategy (Algorithm 1 from Chap. 4). The strategy is then evaluated at run-time taking into consideration these conditions. For the reader's convenience, the aforementioned strategy is repeated in this chapter with the name SCALEFEATURENOISERT and labeled as Algorithm 5. Figure 6.1 also shows the run-time implementation of Pareto-optimal configurations selected off-line. In this case, the configurations would be saved in a Look-Up-Table (LUT) and would be fetched at run-time according to the desired accuracy and cost.

The SCALEFEATURENOISERT strategy in Algorithm 5 consists of a greedy neighborhood search that iteratively reduces the quality of individual features

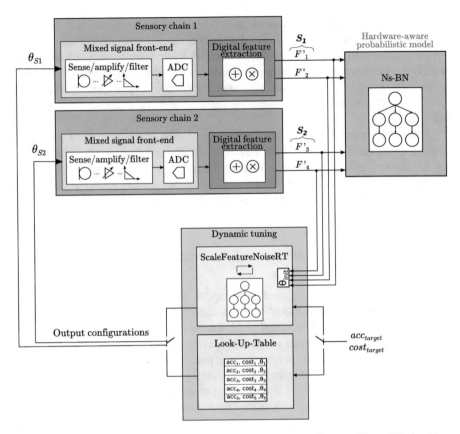

**Fig. 6.1** Dynamic front-end configuration tuning using the SCALEFEATURENOISERT algorithm or a Look-Up-Table

(line 12) in order to decrease cost with minimal accuracy losses (objective function OF in line 15).

In Chap. 4, this procedure was assumed to be performed off-line, resulting in a fixed set of Pareto-optimal feature-noise configurations. At application time, the user could select, from among these noise configurations, the one that would meet the desired accuracy versus cost operation.

To account for dynamic conditions considered in this section, this algorithm is instead evaluated at run-time, with knowledge of the device's current state, which can be provided as one of the inputs $\theta_{init}$ (see Fig. 6.1). Recall that the goal of this algorithm is to find the locally optimal configuration of the sensor front-end $\theta^*$ that, for a given cost, results in the highest accuracy and vice versa. This run-time implementation of the algorithm then assumes a closed loop architecture, like the one depicted in Fig. 6.1, which enables the evaluation of the selection algorithm based on the current state of the device and the current cost and accuracy requirements ($acc_{target}, cost_{target}$).

---

**Algorithm 5:** SCALEFEATURENOISERT(BN, $acc_{target}$, $cost_{target}$, OF, $\theta_{init}$)

---

**Input:** BN: trained model, $acc_{target}$, $cost_{target}$: target accuracy and cost, OF: user defined
  objective function, $\theta_{init}$: initial feature-wise configuration

**Output:** $\langle \mathcal{T}, \mathbf{acc}, \mathbf{cost} \rangle$: selected feature configurations, their accuracy and cost

**1** $\theta_{select} \leftarrow \theta_{init}$

**2** $\mathcal{Q} \leftarrow$ SampleBN($BN$)

**3** $acc_{select} \leftarrow$ Acc(BN, $\theta_{select}$, $\mathcal{Q}$)

**4** $cost_{select} \leftarrow$ CostF($\theta_{select}$)

**5** $\mathcal{T} \leftarrow \theta_{select}$, $\langle \mathbf{acc}, \mathbf{cost} \rangle \leftarrow \langle acc_{select}, cost_{select} \rangle$

**6** **while** $acc_{select} \leq acc_{target} \wedge cost_{select} \geq cost_{target}$ **do**

**7** $\quad$ $\mathcal{Q} \leftarrow$ SampleBN($BN$)

**8** $\quad$ **foreach** $\theta_i^{(v_i)} \in \theta_{select}$ **do**

**9** $\quad\quad$ **if** $\theta_i^{(v_i)} ==$ max($\Theta_i$) **then**

**10** $\quad\quad\quad$ $\theta_{cand,i} \leftarrow \theta_{select}$

**11** $\quad\quad$ **else**

**12** $\quad\quad\quad$ $\theta_{cand,i} \leftarrow \theta_{select} \setminus \theta_i^{(v_i)} \wedge \theta_i^{(v_i+1)}$

**13** $\quad\quad\quad$ $acc_{cand,i} \leftarrow$ Acc(BN, $\theta_{cand,i}$, $\mathcal{Q}$)

**14** $\quad\quad\quad$ $cost_{cand,i} \leftarrow$ CostF($\theta_{cand,i}$)

**15** $\quad$ $\theta_{select} \leftarrow \underset{\theta \in \theta_{cand}}{\text{argminOF}}(acc_{cand}, cost_{cand})$

**16** $\quad$ $acc_{select} \leftarrow$ Acc(BN, $\theta_{select}$, $\mathcal{Q}$)

**17** $\quad$ $cost_{select} \leftarrow$ CostF($\theta_{select}$)

**18** $\quad$ $\mathcal{T}$.insert($\theta_{select}$), $\mathbf{acc}$.insert($acc_{select}$), $\mathbf{cost}$.insert($cost_{select}$)

**19** $\quad$ **if** $\forall i : \theta_i^{(v_i)} ==$ max($\Theta_i$) **then**

**20** $\quad\quad$ **break**

**21** **return** $\langle \mathcal{T}, \mathbf{acc}, \mathbf{cost} \rangle$

---

However, Algorithm 5 contemplates several accuracy estimations (one per candidate, as in line 13) at every iteration (loop in line 8). For the off-line implementation this would be evaluated on a validation set $\mathcal{F}_{val}$. In the run-time case, storing and fetching a validation set for every evaluation of $acc$ would be very inefficient (recall from Sect. 3.4 that memory transactions are costly). Therefore, Algorithm 5 relies solely on the model to select the locally optimal feature configurations. However, analytically computing its expected accuracy can be intractable. The strategy relies instead on a forward sampling process of the ns-BN that simulates a dataset $\mathcal{Q}$ (e.g. line 7 of the algorithm) on which the necessary accuracy estimations are performed (e.g. line 13). This approach serves the goal of identifying what feature-wise noise configuration locally maximizes the desired objective function OF, without having to compute accuracy on an external validation dataset $\mathcal{F}_{valid}$.

Figure 6.2 illustrates the forward sampling process on a four feature ns-BN. The relation Pr($\mathbf{F}'|\mathbf{F}$) is parametrized according to the current state of the device ($\theta_{init}$) for the initial accuracy calculation (line 3 in Algorithm 5) or by the current candidate $\theta_{cand}$ under consideration (line 11) for the accuracy calculation within the iterative feature-noise selection.

**Fig. 6.2** Sampling of $\mathcal{Q}$:
Each data vector $\mathbf{q_i}$ contains
sampled feature values $\mathbf{q}_s$ and
the class it was sampled from
$c_s$

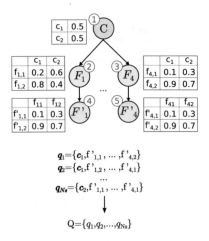

$$q_1 = \{c_1, f'_{1,1}, \cdots, f'_{4,2}\}$$
$$q_2 = \{c_1, f'_{1,2}, \cdots, f'_{4,1}\}$$
$$\cdots$$
$$q_{N_s} = \{c_2, f'_{1,1}, \cdots, f'_{4,1}\}$$

$$\downarrow$$

$$Q = \{q_1, q_2, \ldots, q_{N_s}\}$$

A total number of $N_S$ instances are drawn from the Bayesian network, producing the simulated dataset $\mathcal{Q} = \{\mathbf{q}_1, \ldots, \mathbf{q}_{N_S}\}$ with $\mathbf{q} = \{cs, q_1, \ldots, q_d\}$, where $cs$ is sampled from the class node and the instances $q$ are sampled from the noisy feature nodes. Algorithm 5 then estimates accuracy on this sampled dataset for the evaluation of the objective function OF. This process is repeated for each candidate feature-configuration considered by Algorithm 5, until the desired accuracy and cost are reached.

The following experimental section illustrates situations where sensors fail or become unavailable and compares the performance of the Pareto-optimal set when selected at run-time, and when selected off-line and saved in a Look-Up-Table (LUT) for run-time fetching.

## 6.1.2   Experiments: Robustness to Missing Features of ns-BN

The experiments in this Section consider once again the human activity recognition dataset from [7]. It undergoes the same pre-processing steps described in Sect. 4.4.1. In particular, a 37 feature subset is used for these experiments, and the results are generated by conducting a 5-fold cross validation over 5 trials (each which a new set of folds). Of the 37 features, 8 correspond to the x-accelerometer, 9 to the y-accelerometer, 7 to the z-accelerometer, and the rest to the gyroscope.

The first experiment aims to find an appropriate size ($N_S$) for the sampled dataset $\mathcal{Q}$, which is used for the subsequent run-time evaluation of Algorithm 5, as shown in Fig. 6.3. The blue dotted line shows the off-line results first presented in Sect. 4.4.1, while the orange, yellow, and purple lines show the results of Algorithm 5 when evaluating accuracy on a sampled dataset of sizes equal to $IN, IN/5$, and $IN/10$, where $IN$ is the total number of instances available in the dataset (see Table 4.1). Overall, accuracy deviates from the dataset-based results by less than 2%. Naturally,

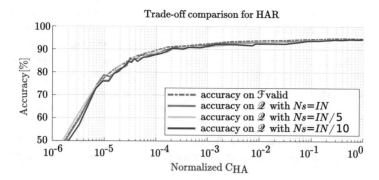

**Fig. 6.3** Accuracy comparison: evaluated on the validation set $\mathcal{F}_{valid}$, or on the sampled set $\mathcal{Q}$ with different sizes

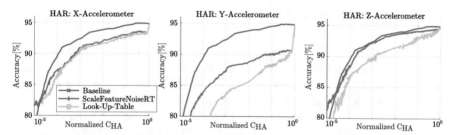

**Fig. 6.4** Performance comparison for the SCALEFEATURENOISERT dynamic tuning approach and the Look-Up-Table approach during sensor stream malfunction

the smaller the sample size, the larger the standard deviation across the different trials will be. The following experiments use a sample size equal to $N_S = IN/5$ for the evaluation of Algorithm 5, which is the same as the size of one of the folds. The motivation behind this choice is that such a sample size will require less iterations for the evaluation of the algorithm, and thus the overhead from sampling is significantly reduced.

Figure 6.4 illustrates sensor malfunction scenarios for the human activity recognition dataset. It compares the results from evaluating Algorithm 5 (SCALEFEATURENOISERT) at run-time with the sampling-based accuracy estimation (red) to the results from fetching off-line selected configurations from the LUT, as shown in Fig. 6.1. Sensor failing takes place by making all the features corresponding to the failing sensor unavailable for observation, and thus marginalizing them during inference. The blue curve shows the results with fully functional sensors and serves as a baseline. Whereas the SCALEFEATURENOISERT strategy is able to, at run-time, pick a feature combination that does not require the usage of the failed or deactivated sensor, the LUT methodology is bound to the settings selected off-line, which are not optimized to account for this situation.

Consider the experiment where the z-accelerometer becomes unavailable after the first iteration. The on-line methodology (red) is capable of selecting a configuration that achieves almost the same performance as the reference with fully functional sensors (blue). In contrast, the LUT methodology (yellow) experiences an accuracy drop of more than 2% as soon as the sensor becomes unavailable. Similarly, the online SCALEFEATURENOISERT methodology is significantly more robust than the LUT approach when the y-accelerometer is unavailable. The drop in accuracy is overall higher in this experiment because some of the features extracted from this sensor tend to hold a higher correlation with the class variables than in the case of features from the z-accelerometer.

Overall, it can be concluded that the cost versus accuracy performance of the LUT run-time approach degrades more than the SCALEFEATURENOISERT approach.

### 6.1.3  Digital Overhead Costs

Recall from the sensory embedded pipeline introduced in Sect. 2.4 that run-time tuning algorithms are implemented in the digital *dynamic tuning block*, as also shown in Fig. 6.1. Therefore, its cost is determined by the number of elementary operations needed to implement the sampling-based Pareto-optimal extraction algorithm and estimate accuracy on it. As discussed in Sect. 3.5, the tuning of the sensor front-end will typically run at a lower rate than feature extraction and classification. For example, the sensors may have a sampling frequency equal to 1kHz, meaning that the incoming sensory signal must be processed once every 0.001 s. While it may only be necessary to re-tune the quality configuration of the front-end every minute. Thus, the dynamic tuning block will only consume energy every minute, and therefore does not have a significant impact on the drainage of the battery at application time. The following paragraph offers a quantitative analysis of the expected digital overhead from implementing the on-line tuning strategy.

Recall that the run-time evaluation of Algorithm 5 involves the sampling of the ns-BN to extract a simulated dataset $Q$ on which the accuracy required by the OF is evaluated. Considering the necessary memory operations, as well as all the actions that have to be performed to sample the Bayesian network and estimate the expected accuracy, the elementary operation count per iteration of Algorithm 5 depends on the number of features $d$, the (average) cardinality of these features $|F|$, the cardinality of the class variable $c$, and the number of samples $N_S$ in $Q$ and is given as (based on the inference costs discussed in Sect. 3.4): $(d + 1) \times 2N_S$ memory fetches; $(d + 1) \times 2N_S$ register operations; $(2d + 2) \times 2N_S$ multiplications; and $N_S \times (3 + c + d \times (1 + |F|))$ additions.

Table 6.1 shows the overhead of the sampling and expected accuracy estimation required by Algorithm 5 for different sizes of $N_S$. This table presents three different scenarios. In the worst case scenario, the system is first extracting all features at their highest available quality and, over the course of the following minute, the cost needs to be scaled by 5x (while minimizing accuracy losses) by using a sample size $N_S = IN$. In this situation, the overhead cost is 4% of $C_{HA,target}$. In the average and

**Table 6.1** Overhead of running Algorithm 5 every minute

| Scaling scenario | Relative cost |
|---|---|
| 1. Worst: $C_{HA,init}=C_{HA,max}$ to $C_{HA,target}=C_{HA,max}/5$, $N_S=IN$ | 4% |
| 2. Average: $C_{HA,target}=C_{HA,max}/5$ to $C_{HA,target}=C_{HA,max}/10$, $N_S=IN/5$ | 0.2% |
| 3. Best: $C_{HA,init}=C_{HA,max}/10$ to $C_{HA,target}=C_{HA,max}/20$, $N_S=IN/10$ | 0.06% |

best case scenarios, where cost has to be reduced from 5x to 10x, or from 10x to 20x with smaller sample sizes ($N_S$) the overhead is even smaller.

### 6.1.4   Remaining Challenges

In many run-time scenarios, external conditions that might render sensory signals unreliable cannot be predicted or identified. The device might also not be equipped to monitor the state of its own sensors and detect whether they are functioning properly. Moreover, the strategy above addresses a single scalable property of the device (noise-tolerance for feature extraction), while other hardware-aware methods and models may target additional properties, such as model complexity and computational precision. Such is the case of the hardware-aware PSDDs discussed throughout Chap. 5 and the reason why a sequential search strategy was proposed in the exploration of the cost versus accuracy trade-off space. Scaling all properties within the same local search (like the one provided by Algorithm 5) would imply an exponential number of candidates at each iteration, rendering the search intractable for large input spaces.

Section 6.2 proposes a run-time strategy that addresses the aforementioned challenges and leverages the hardware-aware PSDD learning methods discussed in Chap. 5. Furthermore, it proves to be particularly effective at low-cost regions of the trade-off space, and with minimal overhead, which constitutes one of the main concerns for the always-on functionality of embedded applications.

## 6.2   Run-Time Implementation of Hardware-Aware PSDDs: Introduction

As discussed in Chap. 5, PSDDs lend themselves to hardware-aware learning and inference, due to the balance between expressiveness and tractability they strike.

Section 5.3 showed how a variety of device configurations—in terms of, for example, number of sensors, features, and bits—can be mapped to different PSDDs with different levels of complexity and feature spaces and how this, in turn, maps to the cost versus accuracy trade-off space. The following section exploits this hardware-aware mapping to propose a run-time strategy that adapts the complexity of the used model to the difficulty of the current inference task. The strategy relies on a pair of models (one of them is *simple* and the other one is *complex*) and implements a policy that predicts the difficulty of classifying the current sample, switching between the two models accordingly at run-time.

The strategy thus leverages the expressiveness and general discriminative abilities of the complex model for difficult-to-classify samples and exploits the extremely low cost of the simple classifier for easier-to-classify samples. As demonstrated on the Human Activity Recognition benchmark considered throughout this chapter, the proposed strategy is capable of reaching a variety of near-Pareto-optimal operating points while relying only on a pair of models, as opposed to a large array of them. Experiments also show that the strategy is particularly effective at low-cost regions of the trade-off space, where it can reach operating points located beyond the static Pareto-optimal curves derived in the last chapter. Furthermore, it displays superior robustness to missing features from unavailable sensors, as it is able to adapt to such an unexpected situation at run-time, unlike a single model selected off-line.

## 6.3  Model Complexity Switching Strategy

The premise behind this strategy is that each model has a level of difficulty associated with classifying a given input. And this level of difficulty on the sample of interest can be deduced by analyzing the conditional probability and the predicted class outputted by the model. For example, in a multi-class activity recognition task a simple model may often confuse walking for running and may therefore output similar conditional probabilities (e.g., 30% and 40%) on the two activities due to the ambiguity. While a more complex model may be capable of correctly identifying one activity from the other in most instances, predicting them with a high conditional probability (e.g., 80% versus 5%).

An overview of the model complexity switching strategy is shown in Fig. 6.5. The core of the strategy, highlighted in light blue, compares the Bayes-optimal conditional probability of the current classifier $\kappa^{(\tau)}$ to the corresponding threshold $\Omega$ and decides whether it is necessary to switch to the other model. More precisely, a different threshold $\Omega$ is available for each model and predicted class, determining whether to go from the simple to the complex one or vice versa. These thresholds are set at an off-line stage by estimating the accuracy on subsets of data corresponding to different ranges of conditional probabilities and predicted classes, as explained in Sect. 6.3.2. Prior to this, the model selection process is explained in Sect. 6.3.1.

The policy discussed above is enclosed by an outer loop, which considers the temporal aspects of the targeted applications and applies the policy periodically. Specifically, the strategy assumes the availability of a timeseries, where a prediction at a given moment in time can be correlated to the previous ones. For example, in the case of activity recognition, a new sample is available for classification every few milliseconds, but activities usually last for several seconds. Therefore, a number of

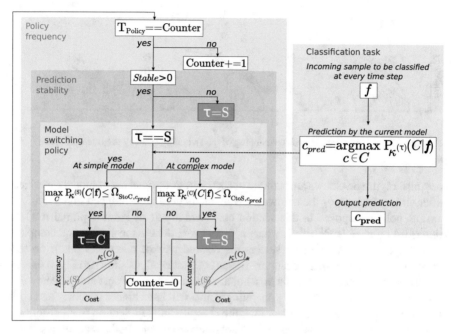

**Fig. 6.5** Overview of the model switching strategy

samples before or after the current one likely belong to the same activity (or class). The strategy therefore applies the policy every $T_{Policy}$ samples (as highlighted in orange), and only as long as the predictions in the past were not stable, which is managed by the variable *Stable* (highlighted in green). That is, the policy is applied if the $T_{Policy}$ predictions before the current one were not the same. Otherwise, the strategy deems the current classification task easy and forces the use of the simple model ($\tau = S$). The motivation behind such a scheme is discussed in Sect. 6.3.3.

## 6.3.1 Model Selection

The model switching policy relies on pairs of models extracted from the Pareto-optimal fronts generated with the hardware-aware strategies of Chap. 5. Recall from Sect. 5.3 that the hardware-aware trade-off extraction procedure outputs a set of Pareto-optimal system configurations $\boldsymbol{\zeta^*}=\{\{\kappa_i^*, \mathbf{F}^*_i, \mathbf{S}^*_i, nb_i^*\}_{i=1:p}\}$, where $\kappa$ denotes the model (PSDD in the case of Chap. 5), $\mathbf{F}$ refers to the available feature subset, $\mathbf{S}$ to the sensor subset, and $nb$ to the number of bits used to represent parameters and perform the arithmetic required by inference. Each Pareto-optimal model $\kappa_i^*$ in $\boldsymbol{\zeta^*}$ is encoded in terms of the corresponding system configurations: it represents a distribution over the variable set $\mathbf{F}^*_i$, which is procured by the sensor subset $\mathbf{S}^*_i$, and the sums, products, and parameters that comprise it are encoded with

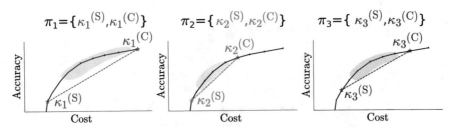

**Fig. 6.6** Model pair selection and the mapping of the switching strategy the trade-off space

$nb_i^*$ bits. Figure 6.9 shows an example of such Pareto-optimal set of models, attained after different stages of the hardware-aware methodology introduced in Sect. 5.3.

The policy proposed in this section considers a pair of Pareto-optimal models described by $\pi = \{\kappa^{(S)}, \kappa^{(C)}\}$, where $\kappa^{(S)}$ and $\kappa^{(C)}$ denote the simple and complex model, respectively. The user selects the desired pair of classifiers based on the region of interest within the accuracy versus cost trade-off, as shown in Fig. 6.6. The accuracy and cost of the models in $\pi$ bounds the accuracy and cost attainable by the switching strategy: the resulting operating points will have higher accuracy and cost than the simple model, and lower than the complex model. Section 6.3.2 describes how these accuracy versus cost regions are procured by the model switching policy.

Recall that one of the goals of the model switching strategy is to be able to reach the regions that constitute Pareto-optimal performance (black lines in Fig. 6.6) at run-time while only storing two models. The selection of the model pairs takes this into consideration, as well as the priority over lower cost or higher accuracy. For example, in Fig. 6.6, pair $\pi_1$ prioritizes higher accuracy and pair $\pi_2$ lower cost. Therefore, both pairs can be available for the user to select when deploying this strategy at application time, since it covers most of the Pareto curve. An example of this will be discussed in the experiments of Sect. 6.3.4.

## 6.3.2   Model Switching Policy

The switching policy highlighted in blue in Fig. 6.5 determines, at run-time, whether to use the simple model $\kappa^{(\tau=S)}$ or the complex one $\kappa^{(\tau=C)}$. This decision is made by comparing the Bayes-optimal conditional probability of the current model $\max P(C|\mathbf{f})$ on the available instance $\mathbf{f}$, to a threshold $\Omega$, determined by calculating the *estimated accuracy* of the current instance $\mathbf{f}$ as detailed in the following paragraphs.

**Setting the Thresholds**

Recall that, given PSDD $\kappa$, the Bayes-optimal conditional probability of example $\mathbf{f}$ in training set $\mathcal{F}_{\text{train}}$ is given by $\max_{c_{\text{pred}} \in C} \Pr_\kappa (C|\mathbf{f})$, where $c_{\text{pred}}$ is the predicted class value. The policy compares the threshold $\Omega$ with the Bayes-optimal probability of the current model $\kappa^{(\tau)}$ to determine whether to move to a different model. For explanation ease, the Bayes-optimal conditional probability is denoted from now on with $P$ and the discrete[1] values it can take with $p$. The notation for the training dataset will also be shortened to $\mathcal{F}$.

Parameters $\Omega$ aim to determine the range of values of $P$ at which it is necessary to switch to a different model. This is done by estimating the accuracy of the subset within $\mathcal{F}$ that is consistent with each possible combination of predicted class and probability value $(c_{pred}, p)$. In other words, there is an *estimated accuracy* associated with each tuple $(c_{pred}, p)$. This metric is denoted with $\widehat{acc}$ and defined by the following equation:

$$\widehat{acc}(c_{pred}, p) = \frac{\mathcal{F}\#([c_{pred}] \wedge [P <= p] \wedge [\text{TP}])}{\mathcal{F}\#([c_{pred}] \wedge [P <= p])}, \qquad (6.1)$$

where $\mathcal{F}\#$ denotes the number of examples of the training dataset (or the frequency counts) that meet the logical condition in parenthesis. The condition $[c_{pred}]$ denotes instances classified as $c_{pred}$, and $[\text{TP}]$ denotes true positive examples, or correctly classified examples,[2] as shown by the confusion matrices at the top of Fig. 6.7. The condition $[P <= p]$ refers to those examples that were predicted with a Bayes-optimal probability lower or equal than value $p$.

The bottom of Fig. 6.7 shows the $\widehat{acc}$ corresponding to $c_{pred}$ equal to activity 1, activity 3 and activity 6 for two models trained in the human activity recognition dataset. The value of $\widehat{acc}(c_{pred} = 1, p = 0.9)$ for $\kappa^{(C)}$, for example, is equal to 0.7. The confusion matrix corresponding to $\kappa^{(C)}$ shows that the number of true positive examples for this class is 1188. Out of those true positive examples, 163 are predicted with a probability less or equal than 0.9. While there are 79 samples wrongly predicted as class 1, 70 of which have a probability less or equal than 0.9. The value of $\widehat{acc}$ can then be calculated by dividing 163 by 163+70, which equals 0.7.

The switching strategy proposed throughout this section relies on thresholds on the Bayes-optimal probabilities calculated by both models (see the blue block in Fig. 6.5). The process of setting these thresholds takes into consideration the fact that the simple model is less expressive than the complex one. The former is not

---

[1] For this, the values of the Bayes-optimal probability are binned in intervals (e.g., 10 intervals of size 0.1).

[2] Recall that for the multi-value classification in this books we refer to all correctly classified samples as true positive, unlike in the case of binary classification where true positives often refer to correctly classified samples belonging to the positive class.

Confusion matrix on training data for $\kappa^{(S)}$

| | | Predicted class | | | | | |
|---|---|---|---|---|---|---|---|
| | | c=1 | c=2 | c=3 | c=4 | c=5 | c=6 |
| True class | c=1 | $TP_1$=1240 | 0 | 0 | 0 | 49 | 0 |
| | c=2 | 1116 | $TP_2$=3 | 0 | 0 | 38 | 0 |
| | c=3 | 0 | 0 | $TP_3$=1459 | 0 | 0 | 0 |
| | c=4 | 7 | 0 | 0 | $TP_4$=812 | 0 | 483 |
| | c=5 | 447 | 0 | 0 | 0 | $TP_5$=650 | 0 |
| | c=6 | 15 | 0 | 0 | 120 | 0 | $TP_6$=1285 |

Confusion matrix on training data for $\kappa^{(C)}$

| | | Predicted class | | | | | |
|---|---|---|---|---|---|---|---|
| | | c=1 | c=2 | c=3 | c=4 | c=5 | c=6 |
| True class | c=1 | $TP_1$=1188 | 58 | 0 | 0 | 43 | 0 |
| | c=2 | 27 | $TP_2$=1074 | 0 | 0 | 56 | 0 |
| | c=3 | 0 | 1 | $TP_3$=1458 | 0 | 0 | 0 |
| | c=4 | 7 | 5 | 0 | $TP_4$=1063 | 2 | 232 |
| | c=5 | 42 | 71 | 0 | 0 | $TP_5$=984 | 0 |
| | c=6 | 3 | 3 | 0 | 180 | 1 | $TP_6$=1233 |

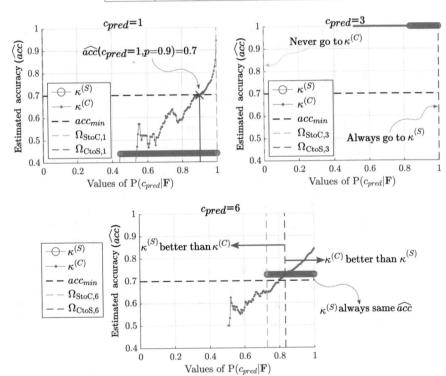

**Fig. 6.7** Top: Confusion matrices for $\kappa^{(S)}$ and $\kappa^{(C)}$ when predicting the samples in $\mathcal{F}_{\text{train}}$. Bottom: estimated accuracy example for three predicted classes and thresholds used by the policy

well equipped to represent all the intricacies found in the training data, and therefore tends to predict several instances corresponding to different observations with the same Bayes-optimal probability (note in Fig. 6.7 that all the lines corresponding to $\kappa^{(S)}$ are flat). While the complex one has better discriminative abilities because the granularity of its Bayes-optimal predictions is higher. Therefore, the policy relies on two different thresholds on the Bayes-optimal probability: one that dictates whether to move from the simple to the complex model ($\Omega_{StoC}$), and another one that drives the converse action ($\Omega_{CtoS}$).

The simple-to-complex parameter $\Omega_{StoC}$ is determined by setting a lower bound on $\widehat{acc}$, denoted with $acc_{min}$, which can be defined by an expert or tuned like a hyperparameter as shown in the experiments. This process of setting the parameter manually therefore takes into consideration the fact that the Bayes-optimal probabilities calculated by this model are not very informative (due to its lack of expressiveness as explained above). The policy will go to a complex model whenever the current sample $\mathbf{f}$, classified with tuple ($c_{pred}$, $p$), has an estimated accuracy lower or equal than $acc_{min}$. According to the example in Fig. 6.7, for $c_{pred} = 6$, this happens when $p <= 0.72$ and thus $\Omega_{StoC,6}$ is set to this value, as shown with a green dotted line in the figure.

Parameter $\Omega_{CtoS}$ drives the policy from the complex to the simple classifier when the latter has a higher expected accuracy. Thus, this parameter is set to the value of $P$ where $\widehat{acc}_{\kappa(S)} == \widehat{acc}_{\kappa(C)}$, such that $\widehat{acc}$ is higher for $\kappa^{(S)}$ than for $\kappa^{(C)}$ for values $< p$. This way of setting $\Omega_{CtoS}$ is justified by the fact that the simple model $\kappa^{(S)}$ tends to predict classes always with the same conditional probability value (due to its lack of expressiveness as discussed above). For example, $\kappa^{(S)}$ always predicts class 6 with the same probability (0.73), and it always has the same estimated accuracy (0.72). Then $\Omega_{CtoS}, 6 = 0.83$ (as shown with the magenta line) because predictions on $\kappa^{(C)}$ with probabilities $<= 0.83$ have a lower estimated accuracy than those of $\kappa^{(S)}$. In other words, when the complex model predicts class 6 with a probability $<= 0.83$, its estimated accuracy is always lower than the estimated accuracy of the simple model. So in this situation it is worth taking the risk of going to the simple model, since the complex model is not good at predicting the current sample (as deemed by the estimated accuracy).

Thresholds equal to 0 or 1 signal situations where the policy must always prefer one model over the other. For example, $\Omega_{CtoS,3} = 1.0$ because $\kappa^{(S)}$ has an $\widehat{acc}$ equal to 1.0 over the full range of probability values. Thus, if the current model is $\kappa^{(C)}$ and its prediction is $c_{pred} = 3$, the policy will always switch down to $\kappa^{(S)}$. Similarly, $\Omega_{StoC,1} = 1.0$ since $\kappa^{(S)}$ never meets the $acc_{min}$ requirement. Thus, whenever class 1 is predicted, the policy will switch to the complex model $\kappa^{(C)}$. On the flip side, $\Omega_{StoC,3} = 0$ because the policy should never go from the simple to the complex model when class 3 is predicted. $\Omega_{CtoS,1} = 0$ because the policy must never switch from the complex to the simple model when predicting class 1.

### 6.3.3   Time Aspects

The model switching policy takes into consideration some of the time-series aspects of the application range of interest.

**Prediction and Policy Frequency**

As shown in Fig. 6.5 and highlighted with purple, the current model $\kappa^{(\tau)}$ classifies each incoming sample $\mathbf{f}$ by predicting the class that maximizes the posterior probability $P(C|\mathbf{f})$. The policy switches to another model (updates $\tau$), if necessary, at time $t$, relying on the current model $\kappa_t^{(\tau)}$ and the probability and prediction available. Then, at time $t + 1$, the incoming sample $\mathbf{f}$ is classified with this new model $\kappa_{t+1}^{(\tau)}$.

Moreover, the policy is applied every $T_{Policy}$ time steps, whenever the prediction in the previous time step is not stable (this is explained in the next subsection), as highlighted with orange and green in Fig. 6.5, respectively. Predictions at $t + 1$ can benefit from a model configured at $t$ due to the time-series properties present in the application range of interest. Similarly, these properties justify the choice for a policy frequency $T_{Policy}$ larger than one time step. Specifically, in most embedded portable applications, the different classes (or e.g. activities) materialize at a lower rate than the sampling frequency of the sensors. Thus, the model selected by the switching strategy attempts to address the needs of the inputs available at time $t + x$.

Note that multiple state-of-the-art strategies focus on learning the duration of the activities (or classes relevant to the application), their sequence, or the transitions between them, such that the most appropriate prediction frequency can be inferred. For example, [11] learns to identify transitions between activities in a home setting, and predicts the most likely activity to take place after the current one. The strategy proposed throughout this section focuses on addressing the dynamic aspects of embedded portable applications, and considers the aforementioned time-series-aware approach as an orthogonal problem. It thus assumes that every incoming sample $\mathbf{f}$ at time $t$ will be predicted by the current classifier $\kappa^{(\tau)}$, even though the strategy is only applied every $t + x$ samples. In addition, some of the dynamic changes in portable applications may render the aforementioned sequence- or transition-aware techniques unstable, such as sensor failure or unexpected behavior of the user (for example, the user decides to perform multiple activities in a random order and transition between them at random times). The strategy proposed in this section is capable of run-time adaptation to such dynamic changes as proven by the experiments in Sect. 6.3.8, where sensors become unavailable.

Taking the previous discussion into consideration, the application frequency $T_{Policy}$ is selected empirically in the experimental section. It is also shown that a wide range of application frequencies can result in beyond-Pareto-optimal performance for a variety of application conditions (different activity-length ranges, performed in random sequences).

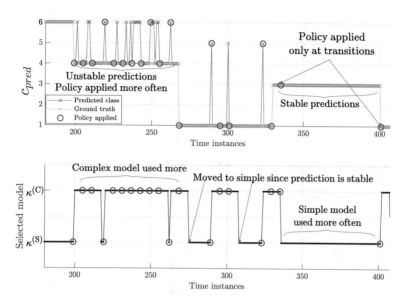

**Fig. 6.8** Example of the strategy implementation

## Prediction Stability

The strategy also takes into consideration the stability of the predictions made in the past to determine whether it is indeed necessary to apply the model switching policy (see section highlighted in green in Fig. 6.5). Specifically, the policy will not be applied if the $T_{\text{Policy}}$ previous samples were predicted to belong to the same class, and the simple model will be used instead. The reasoning behind this is that, if a number of samples were predicted to belong to the same class in the past, it is likely that the current sample also belongs to the same class. Moreover, this stability serves as an indication of a model that is confident. A model that predicts a different class in a very short span of time is likely to be making mistakes and can be considered unreliable.

This is illustrated by Fig. 6.8: the top part of the figure compares the ground truth in green with the predicted class in red and highlights time instances where the model switching policy was applied in blue. In this example, the policy application frequency $T_{\text{Policy}}$ is equal to 5 time instances. Note that, between sample 200 and 250, the model shifts its prediction between class 4 and 6, and that it does so sometimes in the span of two time instances. Throughout this period, the policy is applied every 5 time instances (as often as possible) and the complex model $\kappa^{(C)}$ is selected more frequently (see bottom of figure). On the other hand, the predictions from time 330 to 470 are very stable, and thus the policy is only applied when there are transitions in the prediction. The simple model $\kappa^{(S)}$ is selected for most of these instances, except at the transitions. Indeed, as seen in the confusion matrix and expected prediction examples in Fig. 6.7, class 3 is correctly and confidently predicted by model $\kappa^{(S)}$.

The policy implementation is controlled by a variable called *Stable*, as shown in Fig. 6.5 and computed at time $t$ as follows:

$$
Stable_t = \sum_{i=0}^{T_{\text{Policy}}} [1_{c_{\text{pred},(t-i)} \neq c_{\text{pred},(t-(i+1))}}].
\tag{6.2}
$$

That is, *Stable* is equal to 0 whenever all the predictions from time $t - T_{\text{Policy}}$ to the current time $t$ are exactly the same. When *Stable* is equal to 0, the policy is not implemented and the simple model $\kappa^{(S)}$ is selected.

### 6.3.4   Experiments for Model Switching: Introduction

The model switching strategy is also empirically evaluated on the Human Activity Recognition dataset from [7] and consists on smartphone accelerometer and gyroscope data and six activities.

**Dataset Pre-processing**  Numerical features were discretized using the method in [8] and binarized using a one-hot encoding. The original train-test split was used [7], but a fraction of the test set was set aside for validation purposes for an overall 75%-train, 15%-test, 10-%validation scheme. Similar to the experiments in Sect. 5.4, the dataset underwent an initial feature pre-selection procedure on the binarized train set, using a wrapper feature selection algorithm with a TAN classifier and setting a limit of 55 binary features (this limit was set to constrain the size of the largest available model). The feature set contemplates features extracted from the six available sensory signals (tri-axial accelerometer and tri-axial gyroscope). Finally the test and validation sets are reorganized such that each activity takes place over a period of between 20 and 30, 60 and 70, or 120 and 130 samples and materializes in a random sequence.[3] Furthermore, five different datasets were generated for each activity duration, and the results shown throughout the following sections were averaged over them for statistical precision. This random activity duration and sequence setup allows to verify that the model switching strategy remains functional regardless of activity length and sequence, which other run-time strategies would not be able to cope with since their functionality is bound to sequences learned off-line, as discussed in Sect. 6.3.3. Moreover, this re-ordering of the data aids in demonstrating the impact of setting different values on the hyperparameters discussed in Sect. 6.3.6.

---

[3] The original dataset considered activity duration of between 15 and 30 samples, with a predefined sequence that was repeated periodically.

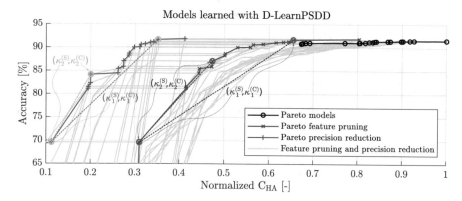

**Fig. 6.9** Training set cost versus accuracy trade-off of the Pareto-optimal sets selected with the techniques in Chap. 5. The models used to evaluate the switching strategy are highlighted in green and magenta

**Model Learning**  The set of models used for the experiments herewith were learned with D-LEARNPSDD [10], as described in Sect. 5.5. More precisely, a model was retained every $N/10$ iterations, where $N$ is the number of iterations needed for convergence. Recall that convergence on D-LEARNPSDD is achieved when the highest validation set accuracy is reached. The training set accuracy and cost for this collection of discriminative-generative models is shown in black in Fig. 6.9. The hardware-aware learning strategy discussed in Sect. 5.3 and introduced in [9] was then applied on this collection of models. The blue line in Fig. 6.9 shows the *feature and sensor set scaling* stage of this methodology, where the feature set **F** is sequentially scaled and the PSDD is pruned accordingly (Algorithm 4), while the red line shows the *precision scaling* stage of the strategy. The gray lines show the intermediate steps taken to reach the final Pareto-optimal settings. Costs are calculated following the description of the experiments in Sect. 5.4 and are normalized according to the largest model learned.

### 6.3.5  Model Selection for Experiments

As discussed in Sect. 6.3.1 the proposed strategy relies on pairs of models extracted from the Pareto-optimal front. Figure 6.9 highlights the models pairs used for experiments in magenta for the *feature and sensor set scaling* case and in green for the *precision scaling* case. The experiments consider two pairs of models per case $\pi^{(SW)} = \{(\kappa_1^{(S)}, \kappa_1^{(C)}), (\kappa_2^{(S)}, \kappa_2^{(C)})\}$. The selected models are those with the highest and lowest training set accuracy per case,[4] as well as the ones with a cost at the midpoint between them.

This choice is motivated by the observation made in Sect. 6.3.1, where the performance of the switching strategy is empirically bounded by the intersection between the pair of models. The two pairs selected above can then span a region close to the original Pareto-optimal front, as illustrated in green and blue ($\kappa_1$ and $\kappa_2$) in Fig. 6.3. As such, the Pareto-optimal trade-off front can be reached with a limited number of models, in the current case three. This will be discussed in more detail throughout Sect. 6.3.7.

### 6.3.6  Hyperparameter Selection

As shown in Fig. 6.5 the strategy's behavior is dictated by two hyperparameters: how often the policy can be evaluated $T_{Policy}$ (which also affects the *Stable* variable); and $acc_{min}$, which sets the threshold $\Omega_{StoC,c_{pred}}$ that guides the policy from the simple to the complex model.

For the experiments in Sect. 6.3.7, hyperparameters $T_{Policy}$ and $acc_{min}$ are selected via a grid search for each activity duration. This selection was made by applying the switching strategy with the available hyperparameter values on the validation set and selecting the cost versus accuracy Pareto-optimal set. The values considered were $T_{Policy} = \{2, 5, 10, 15, 20, 25\}$ and $acc_{min} = \{70\%, 80\%, 90\%\}$ and were selected from and empirical study akin to the one discussed next (see also Appendices B and C for the full list of hyperparameters used for experiments).

The effects of hyperparameters $T_{Policy}$ and $acc_{min}$ are illustrated by Fig. 6.10. Each of the four sub-graphs at the top of the figure shows the performance obtained with different values of $T_{Policy}$ when applying them on different activity-length permutations of the dataset (activity lasts between 20 and 30 samples, between 60 and 70, or between 120 and 130). The lower part of the figure overlaps the results from the different frequencies, for an overall comparison. Moreover, the different types of markers in the upper part of the figure indicate different settings for $acc_{min}$.

Smaller values of $T_{Policy}$ (equal to 2 and 5) allow to reach lower cost regions of the trade-off space. Note that both of these values allow to attain a beyond-Pareto trade-off when compared to the static Pareto-optimal, regardless of activity length. The fact that lower $T_{Policy}$ values lead to lower cost/accuracy regions may seem counter-intuitive, since the policy has the opportunity to be evaluated more often, leading, potentially, to more instances where the complex model is switched to, especially in the case of $acc_{min} = 0.9$. However, recall that the variable *Stable* monitors the changes in predictions in the past over a period of time equal to $T_{Policy}$. Therefore, for a value of $T_{Policy} = 2$, the strategy will tend to move more often to

---

[4]Accuracy was rounded to two significant figures for the selection of these models, since differences smaller than 1% are likely due to numerical effects.

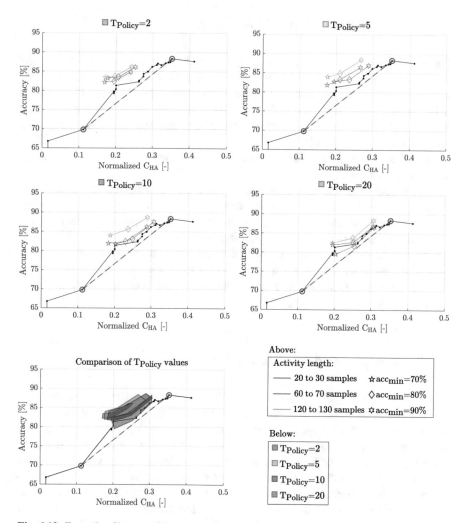

**Fig. 6.10** Example of impact of hyperparameters

the simple model $\kappa^{(S)}$, since it is easy for two consecutive samples to get the same prediction (see Fig. 6.8). For this setting, the strategy tends to execute the policy only in very unstable situations or at transitions between activities. This explains the low-cost/higher-than-Pareto-optimal accuracy this value of $T_{Policy}$ attains. On the other hand, larger values of $T_{Policy}$ (equal to 10 and 20) allow to reach higher accuracy regions of the trade-off space. However, datasets with short activity duration are not capable of coping with the largest value of $T_{Policy}$. In general, values equal to 2 and 5 provide beyond-Pareto results, as confirmed by the hyperparameters selected for the experiments in the following section.

### 6.3.7  Performance of the Model Switching Strategy

Figure 6.11 shows the test set mapping of the Pareto settings extracted in the previous section. As argued in Sect. 6.3.1, the strategy yields operating points located beyond the bound traced by the two models switched between (dotted lines in the figure). Furthermore, the strategy proves to be effective for a variety of activity lengths and sequences (recall that the results herewith are the average over five randomly ordered datasets).

As discussed in the previous section, most of these results consider values of $T_{Policy}$ equal to 2 or 5.

Finally, note that the proposed strategy delivers operating points close to the static Pareto, or better in low-cost regions, and it does so while only relying on pairs of models. At application time, the strategy may be capable of reaching most of the static Pareto-optimal range by tuning the hyperparameters dynamically. For example, if the battery's level is very low the user could opt for $\Omega_{StoC}$ corresponding to low values of $acc_{min}$; or for a higher value when low-cost operation is not a priority. This can be done by, for example, storing three sets of $\Omega_{StoC}$ corresponding to three possible values of $acc_{min}$.

### 6.3.8  Robustness to Missing Features

The switching strategy also proves to be effective when dealing with unexpected run-time changes. Figure 6.12 shows an example of such a situation and compares the performance of the proposed strategy to that of using a single model selected off-line. Specifically, this experiment evaluates the performance degradation resulting from the malfunction of the accelerometer in the y direction. The colored markers show again the evaluation of the strategy under this sensor failure example with the Pareto-optimal hyperparameters selected in Sect. 6.3.6. While the solid lines show the performance of the static Pareto under the same instance of sensor failure. As a reference, the performance of the fully functional sensors is shown with the dotted lines and the faintly colored markers for the static Pareto and the switching strategy, respectively. Note that the solid markers indicating the performance of switching models are always above the solid lines representing the single static Pareto under sensor failure.

It is clear that the switching strategy is capable of coping with this example of sensor failure, as it constantly adjusts to the level of difficulty of the task, which increases when features are not observed. This is true regardless of the activities' length and sequence and the selected hyperparameters.

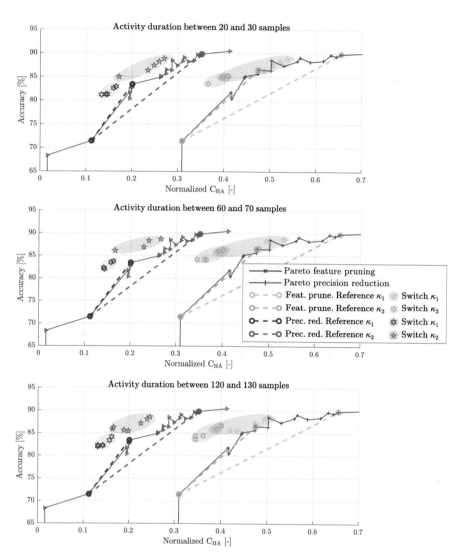

**Fig. 6.11** Results on the test set with Pareto-optimal hyperparameters. The color coding of the reference Pareto curves follows that of Fig. 6.6: the blue one corresponds to the Pareto after *feature and sensor scaling*, and the red one to *precision scaling*. The markers are color coded according to the model pair that produces them, and the dotted line is the intersection between the model pair, setting a bound on the performance of the switching strategy

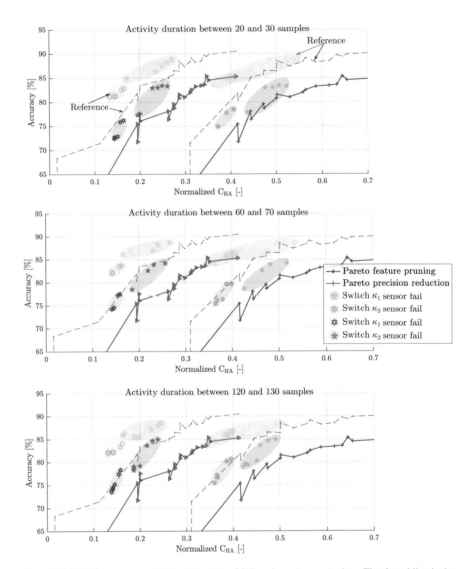

**Fig. 6.12** Robustness to missing features from failure of accelerometer in y. The dotted line is the static Pareto-optimal with fully functional sensors and serves as a reference. The faint markers and enclosing ellipses are the results of the switching strategy with fully functional sensors and also serve as a reference

## 6.4  Related Work

A number of works have focused on addressing the run-time challenges of implementing machine learning algorithms in resource constrained embedded applications.

There is a large body of work catering to the efficient deployment of neural networks that considers selective execution strategies. The work in [12] proposed a type of feed-forward neural network that augments a traditional architecture with controller modules that activate or deactivate subsets of the network depending on the input. Similar to the goal pursued throughout this book, the aforementioned work demonstrates how this selective execution can optimize cost-efficiency trade-offs for different types of tasks. The work in [13] considers a keyword spotting application that exploits such selective execution in a cascaded fashion and further optimizes the models to account for changes in input class distributions, from, for example, situations where environmental noise is very high or very low. Finally, [14] proposes a resource-aware neural network optimization framework that uses different techniques (with selective computing among them) to reconfigure a pre-trained model for practical mobile deployment. However, none of these approaches outline an appropriate strategy to address dynamic run-time changes like the sensor failure discussed throughout this chapter. Some of the properties of the probabilistic models used in this book have proved to be suitable in this regard, such as the fact that they are generative but work well for discriminative tasks and the fact that they encode joint distributions and enable marginal inference.

Closer to the work proposed throughout Sect. 6.3, Inoue proposes a prediction strategy that dynamically determines whether to ensemble more predictions based on confidence levels on their output probability [6]. A number of cost-sensitive learning strategies have also targeted the problem of increasing or decreasing the complexity of feature extraction and inference to trade-off cost and performance. For example, in [5], Xu et al. rely on a sequence of progressively more expressive (and therefore costlier) models, accepting the final prediction that meets the desired confidence level. However, these sequential predictions and ensembling approaches suppose a high overhead that the extreme-edge applications targeted in this book may not be able to support.

Another approach in the field of cost-sensitive learning consists in sequentially deciding whether to observe more features or not, based on the trade-off between the cost of acquiring them and the added performance benefit [1–3]. In [4], Verachtert et al. propose one approach that is capable of efficiently and dynamically controlling such a cost-sensitive forward feature selection. Specifically, it proposes to augment naive Bayes classifiers with "stopping points," which dynamically determine, for each individual query instance, whether more features should be observed before making a decision. However, such approach relies on sequential feature extraction, which might not be suitable to the real-time extreme-edge applications considered in this book. For example, if the aforementioned approach switches from a 5 feature model to a 10 feature model, 5 time steps would be required. Whereas our approach would have a 5 feature model available as well as a 10 feature model available and could switch between them without further delay.

In the realm of resource-efficient run-time probabilistic inference, Vlasselaer et al. propose in [11] a layered computational scheme that selectively activates the use of the state-of-the-art classifiers and uses a probabilistic model to reason about their outputs. Moreover, it relies on a Dynamic Bayesian Network, a type of PGM that can be used to model time dependency, to take into consideration information from the past to determine the present state. The dynamic modeling combined with the hierarchical structure of this work demonstrate cost versus accuracy performance improvements in embedded sensory applications. This work, however, does not exploit scalable hardware or presents a hardware-aware strategy, focusing only on the trade-off of accuracy versus the computational costs. However, the use of a hybrid architecture (using neural networks at a lower level and probabilistic models at a higher level) as well as explicitly modeling time-series aspects are traits of this work that can be taken into consideration to extend the techniques proposed in this book and improve classification accuracy.

## 6.5   Discussion

This chapter addressed two of the challenges plaguing embedded sensory applications: run-time robustness to unavailable information (in the form of failing or unobservable sensors) and significant resource constraints. Section 6.1 showed that the feature-noise tuning strategy proposed in Chap. 4 can be easily adapted to account for this type of run-time changes. Then, Sect. 6.1.2 showed that failure to monitor these changes and adapt to them accordingly can negatively impact the application's performance.

Taking cues from the aforementioned experiments, the strategy proposed in Sect. 6.2 focuses on tuning model complexity at run-time and leverages the different hardware-aware PSDD learning techniques discussed in Chap. 5.

The strategy tunes model complexity according to how difficult it is to classify the currently available instance, which is determined via a threshold set on the model's Bayes-optimal conditional probability and the predicted class. It is argued that the difficulty of classifying a given instance can indirectly provide some information about changing external factors, such as in the case of unavailable sensors. It is indeed proven through the experiments in Sect. 6.3.8 that the switching strategy is capable of working under such failing sensor scenarios to find configurations that prevent significant accuracy degradation. The experiments also show that the proposed strategy is particularly effective at low-cost regions of the trade-off space, making it amenable to embedded applications where resource limitations are very strict and small accuracy losses are acceptable.

As mentioned in the related work in Sect. 6.4, future iterations of the strategies presented herewith can benefit from hybrid and hierarchical architectures that exploit other state-of-the-art models. In addition, many embedded applications, around which this book focuses, provide rich time-series information, which could be included in the model and exploited towards improving cost versus accuracy trade-offs.

In general, there is a vast body of work that looks at how to reason about the performance of classifiers under missing data, and also their reliability under such circumstances [15–17]. Several of the theoretical notions in these works could be incorporated in the strategies proposed throughout this chapter to propose alternative objective functions and policies.

# References

1. Y. Wang, I.I. Hussein, D. Brown, R.S. Erwin, Cost-aware Bayesian sequential decision-making for search and classification. IEEE Transactions on Aerospace and Electronic Systems **48**(3), 2566–2581 (2012)
2. Z. Xu, M. Kusner, K. Weinberger, M. Chen, Cost-sensitive tree of classifiers, in *International Conference on Machine Learning* (2013), pp. 133–141
3. X. Chai, L. Deng, Q. Yang, C.X. Ling, Test-cost sensitive naive bayes classification, in *Fourth IEEE International Conference on Data Mining (ICDM'04)* (IEEE, 2004), pp. 51–58
4. A. Verachtert, H. Blockeel, J. Davis, Dynamic early stopping for naive bayes, in *Proceedings of the Twenty-Fifth International Joint Conference on Artificial Intelligence*, vol. 2016 (AAAI Press, 2016), pp. 2082–2088
5. Z. Xu, M.J. Kusner, K.Q. Weinberger, M. Chen, O. Chapelle, Classifier cascades and trees for minimizing feature evaluation cost. J. Mach. Learn. Res. **15**(1), 2113–2144 (2014)
6. H. Inoue, Adaptive ensemble prediction for deep neural networks based on confidence level, in *The 22nd International Conference on Artificial Intelligence and Statistics* (PMLR, 2019), pp. 1284–1293
7. D. Anguita, A. Ghio, L. Oneto, X. Parra, J.L. Reyes-Ortiz, A public domain dataset for human activity recognition using smartphones, in *ESANN* (2013)
8. U. Fayyad, K. Irani, Multi-interval discretization of continuous-valued attributes for classification learning, in *IJCAI* (1993)
9. L.I. Galindez Olascoaga, W. Meert, N. Shah, M. Verhelst, G. Van den Broeck, Towards hardware-aware tractable learning of probabilistic models, in *Advances in Neural Information Processing Systems 32 (NeurIPS)* (December 2019)
10. L.I. Galindez Olascoaga, W. Meert, N. Shah, G. Van den Broeck, M. Verhelst, Discriminative bias for learning probabilistic sentential decision diagrams, in *Proceedings of the Symposium on Intelligent Data Analysis (IDA)* (April 2020)
11. J. Vlasselaer, W. Meert, M. Verhelst, Towards resource-efficient classifiers for always-on monitoring, in *Joint European Conference on Machine Learning and Knowledge Discovery in Databases* (Springer, 2018), pp. 305–321
12. L. Liu, J. Deng, Dynamic deep neural networks: Optimizing accuracy-efficiency trade-offs by selective execution, in *Thirty-Second AAAI Conference on Artificial Intelligence* (2018)
13. J. Giraldo, C. O'Connor, M. Verhelst, Efficient keyword spotting through hardware-aware conditional execution of deep neural networks, in *2019 IEEE/ACS 16th International Conference on Computer Systems and Applications (AICCSA)* (IEEE, 2019), pp. 1–8
14. Z. Xu, F. Yu, C. Liu, X. Chen, Reform: Static and dynamic resource-aware DNN reconfiguration framework for mobile device, in *Proceedings of the 56th Annual Design Automation Conference 2019* (2019), pp. 1–6
15. M. Perello-Nieto, E.S. Telmo De Menezes Filho, M. Kull, P. Flach, Background check: A general technique to build more reliable and versatile classifiers, in *2016 IEEE 16th International Conference on Data Mining (ICDM)* (IEEE, 2016), pp. 1143–1148
16. Y. Choi, G. Van den Broeck, On robust trimming of Bayesian network classifiers, in *Proceedings of the 27th International Joint Conference on Artificial Intelligence (IJCAI)* (Jul. 2018)
17. P. Khosravi, A. Vergari, Y. Choi, Y. Liang, G. Van den Broeck, Handling missing data in decision trees: A probabilistic approach, in *The Art of Learning with Missing Values Workshop at ICML (Artemiss)* (July 2020)

# Chapter 7
# Conclusions

The device ubiquity of today's world is undeniably the product of a series of technological breakthroughs, such as the persistent development of semiconductor technology, and the algorithmic innovations stemming from decades-old knowledge in fundamental fields like mathematics, statistics, and probability theory.

In much the same way, future technological paradigms will be built on the foundations laid out by what is currently considered state-of-the-art. Moreover, they will be driven by the demands and interests of today's society.

The shift from cloud-centric to edge computing is of particular interest for machine learning implementations due to its compelling and consequential growth. Clever hardware designs and creative resource-efficient algorithmic strategies have already brought machine learning realizations, most of them relying on DNNs, to the extreme edge. But some of the remaining challenges call for capabilities that DNNs are not equipped with and that can be addressed with other approaches, such as probabilistic models.

The overarching goal of this book was to optimally trade off task-level performance for resource consumption savings, in an attempt to enable the implementation of probabilistic models in resource-constrained extreme-edge devices. This was achieved by endowing the models with hardware-awareness, which holistically addressed the properties of multiple stages of the embedded sensing pipeline. Several traits of probabilistic models were exploited to reach this goal, such as robustness to missing data, expert knowledge encoding capabilities, and uncertainty representation.

The first part of this book focused on the feature extraction portion of the embedded sensing pipeline and showed how the relationship between hardware-cost and feature quality can be exploited to balance cost and accuracy with the aid of a hardware-aware Bayesian network classifier. The second part of this book demonstrated that it is possible to derive system-wide costs in an automated way, and how this can also be utilized towards the target cost versus accuracy balance. The closing chapter proposed run-time strategies that utilize the aforementioned techniques to react to dynamic operating conditions and constraints, such as sensor failure and reduced battery availability.

This chapter provides an overview of the techniques and strategies proposed in this book and identifies its main contributions to the field of resource-aware machine learning in Sect. 7.1. It then reflects upon these contributions, discussing their limitations and remaining challenges, as well as the opportunities that they open up for future research in Sect. 7.2. Finally, Sect. 7.3 closes with a reflection on what role this book plays towards the pursuit of the implementation of probabilistic models at the extreme edge.

## 7.1   Overview and Contributions

The main contributions of this book can be summarized as follows:

**Extending Probabilistic Models with Awareness of Feature Noise** Chapter 4 introduced the noise-scalable Bayesian network classifier, which represents the effects of diverse sources of hardware noise in the quality of the extracted feature. This model was then deployed to determine the level of noise-per-feature that, for a given cost, will result in the highest accuracy or vice versa. The noise-scalable Bayesian network classifier was also deployed within a closed loop in Chap. 6, where the locally optimal noise-per-feature is determined at run-time, and demonstrates to be robust to several instances of sensor failure.

**Extending Probabilistic Models with Awareness of Hardware-Cost** Chapter 5 proposed a methodology that relies on the properties of Probabilistic Sentential Decision Diagrams (PSDDs) to map the energy consumption contributions of the different stages of the embedded sensing/inference pipeline to a hardware-aware cost metric. This chapter then proposed a sequential search strategy that extracts the locally optimal system-wide configuration that, for a given energy-cost, will provide the highest accuracy or vice versa. Chapter 6 then demonstrated how these models can be deployed in a dynamic scenario, proposing a strategy that switches among them according to the difficulty of the task. This strategy achieved near-Pareto-optimal accuracy versus cost performance and proved to be particularly effective at low-cost regions, making it amenable to embedded scenarios where battery life is limited.

**Enabling Deployment in Unreliable Extreme-Edge Nodes** Chapter 5 also proposed to augment the LEARNPSDD algorithm with a discriminative bias, which forces the feature-class relation always to be represented in the learned model. This enabled the learned models to consistently achieve a classification accuracy higher or equal than that of a naive Bayes classifier while remaining robust to missing features. This model learning strategy was exploited in the experiments evaluating the model switching strategy mentioned in the previous point, proving to be robust against failing sensors.

**Showing Practical Feasibility in Several Use Cases** All the contributions mentioned above were evaluated on a human activity recognition dataset, which constitutes a relevant benchmark relating to the application range of interest. The experiments throughout this book were also benchmarked on various publicly available datasets, corresponding to applications ranging from robotic navigation to voice activity detection. Finally, the proposed techniques proved to be adaptable to different types of hardware configurations (e.g. different types of sensors and features) and different noise sources (analog, digital, or mixed signal), proving that the concept of hardware-awareness applied to probabilistic models can cater to a very wide range of device and hardware realizations.

## 7.2 Suggestions for Future Work

Overall, this book has proven that making probabilistic models aware of the properties of the hardware that hosts them can enable them to function with reduced resource consumption and limited accuracy losses. This goal is broadly applicable to any resource-constrained system. Still, this book focuses on extreme-edge IoT nodes because said resource constraints constitute a fundamental limit towards their intelligent capabilities. This book also demonstrates that application-driven challenges relating to robustness and dynamic changes add another layer of difficulty towards this goal and that hardware-aware probabilistic models constitute a feasible solution to address these challenges.

Suggestions for future work build on the contributions listed above, aiming to widen their scope and to incorporate alternative approaches:

**Other Resources and Workloads** The proofs of concept in this book focused on energy consumption since its miss-management can preclude battery-powered devices from functioning at the desired performance. Other resources, such as latency, undeniably have a significant effect on the performance of extreme-edge devices as well. A future work direction consists of extending the strategies proposed throughout this book to optimize the trade-off between classification accuracy and latency. The cost metrics proposed in Chap. 3 extend computational complexity notions to the application realm by considering energy consumption. This can comparably be done for other resources, such as latency. Overall, the solutions proposed in this book followed a well defined set of steps. The process starts with the identification of the hardware blocks implementing the desired task; the definition of cost functions in terms of the resource of interest and the tunable parameters of the system; the proposal of hardware-aware models that represent these tunable parameters and allow to explore the cost versus performance trade-off space; and the proposal of strategies to find the Pareto-optimal trade-off within this space. A future work direction could consist in implementing this pipeline to workloads other than machine learning, or models other than probabilistic ones, such as those found in the fields of robotics or control engineering.

**Hybrid Machine Learning Schemes** The benchmarks and datasets considered throughout this book assume feature engineering, where the developer determines what types of features can be extracted and the process of extracting them is known (e.g. the extraction of statistical features from accelerometer and gyroscope signal). However, some applications and situations may benefit from an automated process of feature extraction, since knowledge about the most suitable feature set is not always available. In that situation, the methods proposed herewith could benefit from hybrid models, where, for example, feature extraction is performed by DNNs, or the predictions in classification are performed by a discriminative model, while the closed loop, or a reasoning layer, is controlled by one of the generative probabilistic models used throughout this book [1]. Such a hybrid approach implies that the hardware-aware cost metrics must factor in the contributions of the additional or alternative models.

**Time-Series Aspects of the Applications** The strategy proposed in Chap. 6 takes into consideration some of the time-series aspects of embedded portable applications. However, it does not represent these aspects explicitly in the models or in the policy. As already suggested in that chapter, learning certain characteristics of the application like the average duration of activities and their length is an orthogonal problem to the one addressed by the proposed strategy, but taking this into consideration could serve to improve the overall performance. A future direction for the techniques proposed in Chap. 4 could rely on other Probabilistic Graphical Models that are equipped to represent the time-series aspects of the applications, such as Dynamic Bayesian Networks, or Hidden Markov Models.

**Hardware-Awareness During Learning** The extraction of the Pareto-optimal trade-off is performed by a hardware-aware local search that is initialized to the most complex model or highest feature quality and attempts to iteratively reduce this complexity or quality. A different approach could contemplate the opposite direction in this procedure, where the learning algorithm is augmented with hardware-awareness. This is particularly promising for the hardware-aware Probabilistic Circuits proposed in Chap. 5, as the learning algorithm already trades off model quality for size. These learning algorithms can then be augmented to optimize a hardware-aware metric that already takes into consideration the task of interest (e.g. classification accuracy), as well as the resource of interest (e.g. energy).

## 7.3  Closing Remark

This book constitutes a small but relevant piece in the puzzle of hardware-algorithm co-optimization of probabilistic models in an effort towards their mainstream adoption in extreme-edge applications. The contributions herewith constitute a proof of concept of how hardware-awareness can help in abstracting the properties and constraints of the devices, allowing innovative algorithm techniques to exploit this knowledge towards optimal always-on functionality.

# Reference

1. L.D. Raedt, S. Dumancic, R. Manhaeve, G. Marra, From statistical relational to neuro-symbolic artificial intelligence, in *IJCAI*, 2020

# Appendix A
# Features Used for Experiments

## A.1 Synthetic Dataset for Digital Scaling ns-BN

This dataset consists of 2000 points sampled from 4 Gaussians $\mathcal{N}(\mathbf{m}_i, \boldsymbol{\sigma}_i)$, $i = \{1, 2, 3, 4\}$, where $\mathbf{m}_1 = \begin{pmatrix} -1.66 \\ -0.33 \\ -0.33 \\ -2.00 \end{pmatrix}$, $\mathbf{m}_2 = \begin{pmatrix} 1.00 \\ 0.5 \\ 1.00 \\ 1.00 \end{pmatrix}$, $\mathbf{m}_3 = \begin{pmatrix} 3.33 \\ 2.00 \\ 0.5 \\ 0.5 \end{pmatrix}$, $\mathbf{m}_4 = \begin{pmatrix} -1.66 \\ -1.43 \\ -0.66 \\ -3.33 \end{pmatrix}$, $\boldsymbol{\sigma}_1 = \begin{pmatrix} 0.80 \\ 1.00 \\ 1.00 \\ 1.00 \end{pmatrix}$, $\boldsymbol{\sigma}_2 = \begin{pmatrix} 0.70 \\ 1.00 \\ 1.00 \\ 1.00 \end{pmatrix}$, $\boldsymbol{\sigma}_3 = \begin{pmatrix} 1.00 \\ 1.00 \\ 1.00 \\ 1.00 \end{pmatrix}$, and $\boldsymbol{\sigma}_4 = \begin{pmatrix} 1.00 \\ 1.00 \\ 1.00 \\ 1.00 \end{pmatrix}$. The Gaussians are defined to have different degrees of overlap in every dimension and have therefore a varying mis-classification risk for different feature combinations. The dataset was quantized at 5, 3, 2, and 1 bits and randomly divided into training and testing sets.

## A.2 Six-Class HAR Classification with ns-BN

2 tBodyAcc-mean()-Y
4 tBodyAcc-std()-X
5 tBodyAcc-std()-Y
9 tBodyAcc-mad()-Z
15 tBodyAcc-min()-Z
24 tBodyAcc-entropy()-Y
25 tBodyAcc-entropy()-Z
26 tBodyAcc-arCoeff()-X,1
1 tBodyAcc-mean()-X
6 tBodyAcc-std()-Z

14 tBodyAcc-min()-Y
22 tBodyAcc-iqr()-Z
27 tBodyAcc-arCoeff()-X,2
28 tBodyAcc-arCoeff()-X,3
34 tBodyAcc-arCoeff()-Z,1
17 tBodyAcc-energy()-X
18 tBodyAcc-energy()-Y
19 tBodyAcc-energy()-Z
20 tBodyAcc-iqr()-X

© The Author(s), under exclusive license to Springer Nature Switzerland AG 2021
L. I. Galindez Olascoaga et al., *Hardware-Aware Probabilistic Machine Learning Models*,
https://doi.org/10.1007/978-3-030-74042-9

7 tBodyAcc-mad()-X

8 tBodyAcc-mad()-Y

10 tBodyAcc-max()-X

12 tBodyAcc-max()-Z

13 tBodyAcc-min()-X

16 tBodyAcc-sma()

5 tBodyAcc-arCoeff()-Z,2

3 tBodyAcc-mean()-Z

11 tBodyAcc-max()-Y

30 tBodyAcc-arCoeff()-Y,1

31 tBodyAcc-arCoeff()-Y,2

32 tBodyAcc-arCoeff()-Y,3

21 tBodyAcc-iqr()-Y

29 tBodyAcc-arCoeff()-X,4

36 tBodyAcc-arCoeff()-Z,3

23 tBodyAcc-entropy()-X

33 tBodyAcc-arCoeff()-Y,4

37 tBodyAcc-arCoeff()-Z,4

# Appendix B
# Full List of Hyperparameters for Feature Pruning

Tables B.1, B.2, B.3, B.4, B.5, B.6, B.7, B.8, B.9, B.10, B.11, B.12, B.13, B.14, B.15, and B.16 list the hyperparameters used for the model switching experiments of Chap. 6, corresponding to the Pareto optimal set after feature pruning. These tables also show the performance of each hyperparameter combination in terms of accuracy and cost.

**Table B.1** $\pi_{\text{fp}}$

| $\Omega_1^*$ | $\Omega_2^*$ | Test-Accuracy | Cost |
|---|---|---|---|
| 0.6 | 0.99 | 0.85 | 0.392 |
| 0.5 | 0.99 | 0.85 | 0.373 |
| 0.5 | 0.90 | 0.84 | 0.344 |
| 0.5 | 0.85 | 0.83 | 0.335 |
| 0.1 | 0.90 | 0.69 | 0.211 |
| 0.1 | 0.80 | 0.69 | 0.209 |

$P_{\text{sw}} = 3T$ (3.84s). Activity duration: 30s

**Table B.2** $\pi_{\text{fp}}$

| $\Omega_1^*$ | $\Omega_2^*$ | Test-Accuracy | Cost |
|---|---|---|---|
| 0.5 | 0.99 | 0.87 | 0.372 |
| 0.5 | 0.95 | 0.86 | 0.357 |
| 0.6 | 0.90 | 0.86 | 0.358 |
| 0.5 | 0.90 | 0.86 | 0.340 |
| 0.2 | 0.99 | 0.77 | 0.280 |
| 0.1 | 0.99 | 0.72 | 0.222 |
| 0.1 | 0.85 | 0.72 | 0.211 |
| 0.1 | 0.80 | 0.71 | 0.209 |

$P_{\text{sw}} = 3T$ (3.84s). Activity duration: 60s

© The Author(s), under exclusive license to Springer Nature Switzerland AG 2021
L. I. Galindez Olascoaga et al., *Hardware-Aware Probabilistic Machine Learning Models*,
https://doi.org/10.1007/978-3-030-74042-9

**Table B.3** $\pi_{\text{fp}}$

| $P_{\text{sw}} = 3T$ (3.84s). Activity duration: 120s | | | |
|---|---|---|---|
| $\Omega_1^*$ | $\Omega_2^*$ | Test-Accuracy | Cost |
| 0.6 | 0.99 | 0.88 | 0.407 |
| 0.5 | 0.99 | 0.87 | 0.390 |
| 0.6 | 0.90 | 0.87 | 0.369 |
| 0.5 | 0.95 | 0.87 | 0.374 |
| 0.6 | 0.85 | 0.86 | 0.343 |
| 0.5 | 0.85 | 0.85 | 0.334 |
| 0.2 | 0.99 | 0.79 | 0.309 |
| 0.1 | 0.99 | 0.73 | 0.231 |
| 0.2 | 0.95 | 0.75 | 0.254 |
| 0.1 | 0.95 | 0.73 | 0.223 |
| 0.2 | 0.90 | 0.74 | 0.238 |
| 0.1 | 0.90 | 0.72 | 0.219 |
| 0.1 | 0.85 | 0.71 | 0.215 |
| 0.1 | 0.80 | 0.71 | 0.210 |

**Table B.4** $\pi_{\text{fp}}$

| $P_{\text{sw}} = 3T$ (3.84s). Activity duration: 200s | | | |
|---|---|---|---|
| $\Omega_1^*$ | $\Omega_2^*$ | Test-Accuracy | Cost |
| 0.6 | 0.99 | 0.89 | 0.410 |
| 0.5 | 0.99 | 0.89 | 0.397 |
| 0.5 | 0.85 | 0.86 | 0.331 |
| 0.5 | 0.80 | 0.86 | 0.331 |
| 0.1 | 0.90 | 0.73 | 0.216 |

**Table B.5** $\pi_{\text{fp}}$

| $P_{\text{sw}} = 5T$ (6.4s). Activity duration: 30s | | | |
|---|---|---|---|
| $\Omega_1^*$ | $\Omega_2^*$ | Test-Accuracy | Cost |
| 0.6 | 0.99 | 0.81 | 0.237 |
| 0.5 | 0.99 | 0.81 | 0.222 |
| 0.6 | 0.85 | 0.80 | 0.211 |
| 0.5 | 0.90 | 0.80 | 0.205 |
| 0.5 | 0.85 | 0.80 | 0.201 |
| 0.1 | 0.80 | 0.66 | 0.126 |

**Table B.6** $\pi_{\text{fp}}$

| $\Omega_1^*$ | $\Omega_2^*$ | Test-Accuracy | Cost |
|---|---|---|---|
| $P_{\text{sw}} = 5T$ (6.4s). Activity duration: 60s | | | |
| 0.6 | 0.99 | 0.85 | 0.236 |
| 0.5 | 0.99 | 0.85 | 0.232 |
| 0.5 | 0.95 | 0.85 | 0.218 |
| 0.6 | 0.85 | 0.85 | 0.212 |
| 0.5 | 0.90 | 0.84 | 0.211 |
| 0.5 | 0.85 | 0.84 | 0.205 |
| 0.2 | 0.99 | 0.75 | 0.173 |
| 0.1 | 0.99 | 0.71 | 0.136 |
| 0.1 | 0.80 | 0.71 | 0.129 |

**Table B.7** $\pi_{\text{fp}}$

| $\Omega_1^*$ | $\Omega_2^*$ | Test-Accuracy | Cost |
|---|---|---|---|
| $P_{\text{sw}} = 5T$ (6.4s). Activity duration: 120s | | | |
| 0.5 | 0.99 | 0.85 | 0.224 |
| 0.6 | 0.95 | 0.85 | 0.220 |
| 0.5 | 0.95 | 0.85 | 0.213 |
| 0.5 | 0.80 | 0.84 | 0.196 |
| 0.5 | 0.85 | 0.84 | 0.190 |
| 0.4 | 0.99 | 0.75 | 0.176 |
| 0.2 | 0.99 | 0.75 | 0.174 |
| 0.1 | 0.99 | 0.72 | 0.140 |
| 0.2 | 0.95 | 0.73 | 0.158 |
| 0.1 | 0.95 | 0.72 | 0.136 |
| 0.1 | 0.90 | 0.71 | 0.130 |
| 0.1 | 0.80 | 0.71 | 0.127 |

**Table B.8** $\pi_{\text{fp}}$

| $\Omega_1^*$ | $\Omega_2^*$ | Test-Accuracy | Cost |
|---|---|---|---|
| $P_{\text{sw}} = 5T$ (6.4s). Activity duration: 200s | | | |
| 0.6 | 0.99 | 0.88 | 0.237 |
| 0.5 | 0.99 | 0.87 | 0.228 |
| 0.6 | 0.85 | 0.87 | 0.202 |
| 0.5 | 0.85 | 0.87 | 0.191 |
| 0.1 | 0.90 | 0.73 | 0.130 |

**Table B.9** $\pi_{\mathrm{fp}}$

| $P_{\mathrm{sw}} = 10T$ (12.8s). Activity duration: 30s | | | |
|---|---|---|---|
| $\Omega_1^*$ | $\Omega_2^*$ | Test-Accuracy | Cost |
| 0.5 | 0.99 | 0.73 | 0.112 |
| 0.6 | 0.85 | 0.73 | 0.109 |
| 0.5 | 0.95 | 0.73 | 0.110 |
| 0.4 | 0.99 | 0.63 | 0.079 |
| 0.1 | 0.99 | 0.62 | 0.070 |
| 0.1 | 0.95 | 0.62 | 0.068 |
| 0.1 | 0.90 | 0.61 | 0.068 |
| 0.1 | 0.80 | 0.61 | 0.065 |

**Table B.10** $\pi_{\mathrm{fp}}$

| $P_{\mathrm{sw}} = 10T$ (12.8s). Activity duration: 60s | | | |
|---|---|---|---|
| $\Omega_1^*$ | $\Omega_2^*$ | Test-Accuracy | Cost |
| 0.5 | 0.95 | 0.80 | 0.108 |
| 0.6 | 0.85 | 0.79 | 0.107 |
| 0.5 | 0.85 | 0.79 | 0.101 |
| 0.2 | 0.99 | 0.68 | 0.075 |
| 0.1 | 0.99 | 0.68 | 0.068 |
| 0.2 | 0.95 | 0.68 | 0.073 |
| 0.1 | 0.95 | 0.67 | 0.067 |
| 0.1 | 0.85 | 0.67 | 0.065 |
| 0.1 | 0.80 | 0.67 | 0.065 |

**Table B.11** $\pi_{\mathrm{fp}}$

| $P_{\mathrm{sw}} = 10T$ (12.8s). Activity duration: 120s | | | |
|---|---|---|---|
| $\Omega_1^*$ | $\Omega_2^*$ | Test-Accuracy | Cost |
| 0.5 | 0.99 | 0.84 | 0.114 |
| 0.5 | 0.95 | 0.83 | 0.111 |
| 0.6 | 0.80 | 0.81 | 0.103 |
| 0.6 | 0.85 | 0.81 | 0.101 |
| 0.5 | 0.80 | 0.81 | 0.101 |
| 0.5 | 0.85 | 0.81 | 0.100 |
| 0.4 | 0.99 | 0.70 | 0.078 |
| 0.2 | 0.99 | 0.70 | 0.078 |
| 0.1 | 0.99 | 0.70 | 0.069 |
| 0.4 | 0.95 | 0.70 | 0.076 |
| 0.2 | 0.90 | 0.69 | 0.070 |
| 0.1 | 0.90 | 0.69 | 0.066 |
| 0.1 | 0.85 | 0.69 | 0.066 |
| 0.1 | 0.80 | 0.69 | 0.066 |

**Table B.12** $\pi_{\mathrm{fp}}$

| $\Omega_1^*$ | $\Omega_2^*$ | Test-Accuracy | Cost |
|---|---|---|---|
| $P_{\mathrm{sw}} = 10T$ (12.8s). Activity duration: 200s | | | |
| 0.5 | 0.99 | 0.86 | 0.116 |
| 0.5 | 0.95 | 0.86 | 0.113 |
| 0.4 | 0.99 | 0.79 | 0.099 |
| 0.1 | 0.99 | 0.77 | 0.088 |
| 0.1 | 0.95 | 0.74 | 0.083 |
| 0.1 | 0.90 | 0.72 | 0.071 |
| 0.1 | 0.80 | 0.72 | 0.066 |

**Table B.13** $\pi_{\mathrm{fp}}$

| $\Omega_1^*$ | $\Omega_2^*$ | Test-Accuracy | Cost |
|---|---|---|---|
| $P_{\mathrm{sw}} = 20T$ (25.6s). Activity duration: 30s | | | |
| 0.5 | 0.85 | 0.59 | 0.055 |
| 0.1 | 0.95 | 0.51 | 0.037 |
| 0.1 | 0.90 | 0.50 | 0.037 |
| 0.1 | 0.80 | 0.49 | 0.034 |

**Table B.14** $\pi_{\mathrm{fp}}$

| $\Omega_1^*$ | $\Omega_2^*$ | Test-Accuracy | Cost |
|---|---|---|---|
| $P_{\mathrm{sw}} = 20T$ (25.6s). Activity duration: 60s | | | |
| 0.5 | 0.85 | 0.69 | 0.052 |
| 0.5 | 0.95 | 0.69 | 0.054 |
| 0.4 | 0.95 | 0.62 | 0.040 |
| 0.1 | 0.95 | 0.62 | 0.036 |
| 0.1 | 0.90 | 0.61 | 0.035 |
| 0.1 | 0.80 | 0.60 | 0.034 |

**Table B.15** $\pi_{\mathrm{fp}}$

| $\Omega_1^*$ | $\Omega_2^*$ | Test-Accuracy | Cost |
|---|---|---|---|
| $P_{\mathrm{sw}} = 20T$ (25.6s). Activity duration: 120s | | | |
| 0.6 | 0.80 | 0.76 | 0.052 |
| 0.6 | 0.85 | 0.77 | 0.055 |
| 0.5 | 0.85 | 0.77 | 0.054 |
| 0.4 | 0.95 | 0.66 | 0.038 |
| 0.2 | 0.85 | 0.65 | 0.036 |
| 0.1 | 0.85 | 0.65 | 0.034 |
| 0.1 | 0.80 | 0.64 | 0.033 |

**Table B.16** $\pi_{\mathrm{fp}}$

| $P_{\mathrm{sw}} = 20T$ (25.6s). Activity duration: 200s | | | |
|---|---|---|---|
| $\Omega_1^*$ | $\Omega_2^*$ | Test-Accuracy | Cost |
| 0.5 | 0.85 | 0.83 | 0.052 |
| 0.1 | 0.99 | 0.77 | 0.043 |
| 0.1 | 0.90 | 0.73 | 0.039 |
| 0.1 | 0.80 | 0.71 | 0.035 |

# Appendix C
# Full List of Hyperparameters for Precision Reduction

Tables C.1, C.2, C.3, C.4, C.5, C.6, C.7, C.8, C.9, C.10, C.11, C.12, C.13, C.14, C.15, and C.16 list the hyperparameters used for the model switching experiments of Chap. 6, corresponding to the Pareto optimal set after precision reduction. These tables also show the performance of each hyperparameter combination in terms of accuracy and cost.

**Table C.1** $\pi_{pr}$

| $P_{sw} = 3T$ (3.84s). Act. duration: 30s | | | |
|---|---|---|---|
| $\Omega_1^*$ | $\Omega_2^*$ | Test-Accuracy | Cost |
| 0.500000 | 0.990000 | 0.854215 | 0.196040 |
| 0.600000 | 0.950000 | 0.846433 | 0.187553 |
| 0.600000 | 0.900000 | 0.842283 | 0.176012 |
| 0.500000 | 0.950000 | 0.841245 | 0.173533 |
| 0.500000 | 0.900000 | 0.837873 | 0.164945 |
| 0.600000 | 0.800000 | 0.835279 | 0.159105 |
| 0.500000 | 0.800000 | 0.831518 | 0.150394 |
| 0.100000 | 0.850000 | 0.694553 | 0.077877 |
| 0.100000 | 0.800000 | 0.691699 | 0.076539 |

**Table C.2** $\pi_{\mathrm{pr}}$

| $P_{\mathrm{sw}} = 3T$ (3.84s). Act. duration: 60s | | | |
|---|---|---|---|
| $\Omega_1^*$ | $\Omega_2^*$ | Test-Accuracy | Cost |
| 0.600000 | 0.950000 | 0.865499 | 0.199927 |
| 0.600000 | 0.900000 | 0.863035 | 0.182434 |
| 0.500000 | 0.950000 | 0.861219 | 0.186992 |
| 0.600000 | 0.850000 | 0.861997 | 0.163801 |
| 0.600000 | 0.800000 | 0.860441 | 0.156779 |
| 0.500000 | 0.850000 | 0.859274 | 0.154478 |
| 0.500000 | 0.800000 | 0.856939 | 0.151064 |
| 0.200000 | 0.990000 | 0.769650 | 0.133859 |
| 0.100000 | 0.990000 | 0.724773 | 0.086200 |
| 0.100000 | 0.950000 | 0.721920 | 0.080070 |
| 0.100000 | 0.800000 | 0.719196 | 0.076931 |

**Table C.3** $\pi_{\mathrm{pr}}$

| $P_{\mathrm{sw}} = 3T$ (3.84s). Act. duration: 120s | | | |
|---|---|---|---|
| $\Omega_1^*$ | $\Omega_2^*$ | Test-Accuracy | Cost |
| 0.500000 | 0.990000 | 0.878470 | 0.206507 |
| 0.600000 | 0.950000 | 0.869909 | 0.197633 |
| 0.500000 | 0.950000 | 0.862646 | 0.179435 |
| 0.600000 | 0.900000 | 0.863943 | 0.180252 |
| 0.500000 | 0.900000 | 0.859274 | 0.165355 |
| 0.500000 | 0.850000 | 0.854345 | 0.151346 |
| 0.500000 | 0.800000 | 0.853178 | 0.146791 |
| 0.300000 | 0.990000 | 0.760700 | 0.130517 |
| 0.200000 | 0.990000 | 0.760700 | 0.127868 |
| 0.100000 | 0.990000 | 0.724384 | 0.085325 |
| 0.200000 | 0.950000 | 0.742672 | 0.101639 |
| 0.200000 | 0.900000 | 0.742672 | 0.097102 |
| 0.100000 | 0.950000 | 0.721271 | 0.083588 |
| 0.100000 | 0.900000 | 0.718936 | 0.082651 |
| 0.100000 | 0.850000 | 0.710895 | 0.078535 |
| 0.100000 | 0.800000 | 0.713230 | 0.076788 |

**Table C.4** $\pi_{\mathrm{pr}}$

| $P_{\mathrm{sw}} = 3T$ (3.84s). Act. duration: 200s | | | |
|---|---|---|---|
| $\Omega_1^*$ | $\Omega_2^*$ | Test-Accuracy | Cost |
| 0.500000 | 0.990000 | 0.890013 | 0.207150 |
| 0.600000 | 0.900000 | 0.873411 | 0.191921 |
| 0.500000 | 0.900000 | 0.871466 | 0.174693 |
| 0.500000 | 0.800000 | 0.862905 | 0.152054 |
| 0.100000 | 0.990000 | 0.743320 | 0.106462 |

**Table C.5** $\pi_{\mathrm{pr}}$

| $P_{\mathrm{sw}} = 5T$ (6.4s). Act. duration: 30s | | | |
|---|---|---|---|
| $\Omega_1^*$ | $\Omega_2^*$ | Test-Accuracy | Cost |
| 0.600000 | 0.990000 | 0.820493 | 0.125537 |
| 0.500000 | 0.990000 | 0.813230 | 0.114044 |
| 0.600000 | 0.950000 | 0.814267 | 0.115805 |
| 0.600000 | 0.850000 | 0.805837 | 0.106564 |
| 0.600000 | 0.800000 | 0.807782 | 0.102528 |
| 0.500000 | 0.800000 | 0.803891 | 0.093281 |
| 0.100000 | 0.800000 | 0.663035 | 0.046472 |

**Table C.6** $\pi_{\mathrm{pr}}$

| $P_{\mathrm{sw}} = 5T$ (6.4s). Act. duration: 60s | | | |
|---|---|---|---|
| $\Omega_1^*$ | $\Omega_2^*$ | Test-Accuracy | Cost |
| 0.500000 | 0.990000 | 0.848249 | 0.120517 |
| 0.500000 | 0.950000 | 0.843320 | 0.102484 |
| 0.500000 | 0.800000 | 0.839559 | 0.090288 |
| 0.100000 | 0.990000 | 0.710117 | 0.052351 |
| 0.100000 | 0.950000 | 0.710765 | 0.049553 |
| 0.100000 | 0.850000 | 0.705707 | 0.047189 |

**Table C.7** $\pi_{\mathrm{pr}}$

| $P_{\mathrm{sw}} = 5T$ (6.4s). Act. duration: 120s | | | |
|---|---|---|---|
| $\Omega_1^*$ | $\Omega_2^*$ | Test-Accuracy | Cost |
| 0.500000 | 0.990000 | 0.861349 | 0.120155 |
| 0.500000 | 0.900000 | 0.845136 | 0.096177 |
| 0.600000 | 0.850000 | 0.850065 | 0.101569 |
| 0.600000 | 0.800000 | 0.852270 | 0.100070 |
| 0.500000 | 0.800000 | 0.844747 | 0.092696 |
| 0.300000 | 0.990000 | 0.735538 | 0.078358 |
| 0.200000 | 0.990000 | 0.736187 | 0.077968 |
| 0.100000 | 0.990000 | 0.724254 | 0.057041 |
| 0.200000 | 0.950000 | 0.730091 | 0.066431 |
| 0.100000 | 0.950000 | 0.718158 | 0.051791 |
| 0.100000 | 0.900000 | 0.716861 | 0.050393 |
| 0.300000 | 0.800000 | 0.709079 | 0.053012 |
| 0.200000 | 0.850000 | 0.708431 | 0.053477 |
| 0.100000 | 0.850000 | 0.703243 | 0.046948 |
| 0.100000 | 0.800000 | 0.703243 | 0.046636 |

**Table C.8** $\pi_{pr}$

| $P_{sw} = 5T$ (6.4s). Act. duration: 200s | | | |
|---|---|---|---|
| $\Omega_1^*$ | $\Omega_2^*$ | Test-Accuracy | Cost |
| 0.600000 | 0.990000 | 0.879507 | 0.129505 |
| 0.500000 | 0.990000 | 0.879507 | 0.119418 |
| 0.600000 | 0.950000 | 0.868872 | 0.115373 |
| 0.600000 | 0.800000 | 0.871725 | 0.099744 |
| 0.500000 | 0.800000 | 0.866537 | 0.089770 |
| 0.100000 | 0.990000 | 0.748119 | 0.067193 |

**Table C.9** $\pi_{pr}$

| $P_{sw} = 10T$ (12.8s). Act. duration: 30s | | | |
|---|---|---|---|
| $\Omega_1^*$ | $\Omega_2^*$ | Test-Accuracy | Cost |
| 0.600000 | 0.990000 | 0.738781 | 0.064760 |
| 0.500000 | 0.990000 | 0.734371 | 0.057776 |
| 0.600000 | 0.900000 | 0.739040 | 0.055455 |
| 0.600000 | 0.800000 | 0.731907 | 0.051159 |
| 0.500000 | 0.800000 | 0.725032 | 0.047642 |
| 0.100000 | 0.990000 | 0.633333 | 0.033183 |
| 0.100000 | 0.800000 | 0.611414 | 0.026221 |

**Table C.10** $\pi_{pr}$

| $P_{sw} = 10T$ (12.8s). Act. duration: 60s | | | |
|---|---|---|---|
| $\Omega_1^*$ | $\Omega_2^*$ | Test-Accuracy | Cost |
| 0.600000 | 0.990000 | 0.800778 | 0.061525 |
| 0.500000 | 0.990000 | 0.800000 | 0.056725 |
| 0.600000 | 0.800000 | 0.791180 | 0.048069 |
| 0.500000 | 0.800000 | 0.793774 | 0.045697 |
| 0.100000 | 0.990000 | 0.694812 | 0.028482 |
| 0.100000 | 0.950000 | 0.675616 | 0.026734 |
| 0.100000 | 0.800000 | 0.668742 | 0.024058 |

**Table C.11** $\pi_{pr}$

| $P_{sw} = 10T$ (12.8s). Act. duration: 120s | | | |
|---|---|---|---|
| $\Omega_1^*$ | $\Omega_2^*$ | Test-Accuracy | Cost |
| 0.500000 | 0.990000 | 0.832425 | 0.059955 |
| 0.600000 | 0.950000 | 0.834501 | 0.058602 |
| 0.500000 | 0.950000 | 0.829313 | 0.054963 |
| 0.600000 | 0.800000 | 0.809728 | 0.047774 |
| 0.500000 | 0.800000 | 0.808431 | 0.045837 |
| 0.200000 | 0.990000 | 0.701038 | 0.033054 |
| 0.100000 | 0.990000 | 0.702075 | 0.029227 |
| 0.100000 | 0.950000 | 0.699481 | 0.028338 |
| 0.100000 | 0.900000 | 0.693515 | 0.026421 |
| 0.100000 | 0.850000 | 0.690921 | 0.025690 |
| 0.100000 | 0.800000 | 0.690921 | 0.025507 |

**Table C.12** $\pi_{pr}$

| $P_{sw} = 10T$ (12.8s). Act. duration: 200s | | | |
|---|---|---|---|
| $\Omega_1^*$ | $\Omega_2^*$ | Test-Accuracy | Cost |
| '0.600000 | 0.990000 | 0.864202 | 0.064551 |
| 0.600000 | 0.800000 | 0.853307 | 0.052053 |
| 0.500000 | 0.800000 | 0.852789 | 0.046610 |
| 0.100000 | 0.990000 | 0.771336 | 0.040272 |
| 0.100000 | 0.800000 | 0.712451 | 0.024646 |

**Table C.13** $\pi_{pr}$

| $P_{sw} = 20T$ (25.6s). Act. duration: 30s | | | |
|---|---|---|---|
| $\Omega_1^*$ | $\Omega_2^*$ | Test-Accuracy | Cost |
| '0.500000 | 0.990000 | 0.592348 | 0.029268 |
| 0.600000 | 0.800000 | 0.593256 | 0.026738 |
| 0.500000 | 0.800000 | 0.587549 | 0.025014 |
| 0.100000 | 0.990000 | 0.527108 | 0.019847 |
| 0.100000 | 0.950000 | 0.517899 | 0.017939 |
| 0.100000 | 0.850000 | 0.501167 | 0.014572 |
| 0.100000 | 0.800000 | 0.498573 | 0.014125 |

**Table C.14** $\pi_{pr}$

| $P_{sw} = 20T$ (25.6s). Act. duration: 60s | | | |
|---|---|---|---|
| $\Omega_1^*$ | $\Omega_2^*$ | Test-Accuracy | Cost |
| '0.600000 | 0.800000 | 0.694034 | 0.025556 |
| 0.500000 | 0.800000 | 0.694034 | 0.025479 |
| 0.100000 | 0.950000 | 0.618677 | 0.014820 |
| 0.100000 | 0.900000 | 0.609857 | 0.014351 |
| 0.100000 | 0.850000 | 0.600778 | 0.013849 |
| 0.100000 | 0.800000 | 0.598184 | 0.013632 |

**Table C.15** $\pi_{pr}$

| $P_{sw} = 20T$ (25.6s). Act. duration: 120s | | | |
|---|---|---|---|
| $\Omega_1^*$ | $\Omega_2^*$ | Test-Accuracy | Cost |
| '0.600000 | 0.800000 | 0.764462 | 0.027276 |
| 0.500000 | 0.800000 | 0.764202 | 0.025936 |
| 0.100000 | 0.900000 | 0.651102 | 0.013745 |
| 0.100000 | 0.950000 | 0.659663 | 0.014152 |
| 0.200000 | 0.800000 | 0.652659 | 0.014698 |
| 0.100000 | 0.800000 | 0.648768 | 0.013062 |

**Table C.16** $\pi_{pr}$

| $P_{sw} = 20T$ (25.6s). Act. duration: 200s | | | |
|---|---|---|---|
| $\Omega_1^*$ | $\Omega_2^*$ | Test-Accuracy | Cost |
| '0.500000 | 0.990000 | 0.850454 | 0.028510 |
| 0.600000 | 0.800000 | 0.838651 | 0.027255 |
| 0.100000 | 0.990000 | 0.755772 | 0.020990 |
| 0.100000 | 0.950000 | 0.734890 | 0.018626 |
| 0.100000 | 0.850000 | 0.710506 | 0.014311 |
| 0.100000 | 0.800000 | 0.713100 | 0.013538 |

# Index

**A**
Analog feature precision scaling, 49
Analog to digital converter (ADC), 45, 46
   quantization noise, 44

**B**
Battery-powered edge devices, 10
Battery-powered mobile devices, 1
Bayesian network classifier, 9
Bayesian networks, 81
   classification, 33
   compactness, 30
   exact inference, 31–32
   parameter learning, 32–33
   performance, 33
   probability distribution, 29
   topological criteria, 31

**C**
Clustering, 6
Conditional mutual information (CMI), 101
Conditional Probability Table (CPT), 29
Cross-validation, 7

**D**
Digital feature precision scaling, 49
Digital scaling ns-BN
   cost and objective function definition,
      69–70
   digital feature extraction block, 68

experimental results, 70, 71
experimental setup, 70
front-end with tunable-precision digital
      feature extraction, 68
noisy and noiseless features, 68
synthetic dataset, 145
Directed Acyclic Graph (DAG), 29
Dynamic tuning block, 118
Dynamic tuning costs, 51–52

**E**
Embedded sensing pipeline, 38, 41, 139
Energy consumption, 41
Extreme-edge computing
   bandwidth intensive applications, 3–4
   computation and storage, 3
   device location and mobility, 4
   low latency, 3
   resource efficiency, 4
Extreme-edge nodes, 140

**F**
False positives (FPs), 7
Feature and sensor set scaling stage, 129
Feature extraction costs
   precision and energy consumption, 47
   relative energy consumption, 47
Feature extraction stage, 11
Feature noise tuning, 57, 58
Feature pruning
   hyperparameters, 147–152

© The Author(s), under exclusive license to Springer Nature Switzerland AG 2021      159
L. I. Galindez Olascoaga et al., *Hardware-Aware Probabilistic Machine Learning Models*,
https://doi.org/10.1007/978-3-030-74042-9

**G**
Geometric models, 8
*Global Datasphere,* 1


**H**
Hardware-aware cost models, 52
    concept, 41
    definition, 42
    energy consumption management, 41
Hardware-aware model, 142
    ns-BN, 55 (*see also* Noise-scalable
            Bayesian Network (ns-BN)
            classifier)
Hardware-aware probabilistic circuits
    communities, 106
    cost calculation, 81
    discriminative model, 107
    dynamic model-complexity scaling
            strategies, 108
    embedded devices, 106
    embedded sensing pipeline, 81
    experiments, 107
    fixed tractability constraint, 106
    learning technique, 107
    opportunities, 108
    pareto-optimal trade-off extraction (*see*
            Pareto-optimal trade-off extraction)
    probabilistic inference, 106
    PSDD (*see* Probabilistic Sentential
            Decision Diagrams (PSDDs))
    scaling inference complexity, 81
    system-wide cost, 85–86
    system-wide hardware-awareness, 81
    techniques, 82, 106
Hardware-aware probabilistic model, 39
Hardware cost, 140–141
Hardware noise, 57, 61
Holistic resource-aware machine learning, 14
Human activity recognition (HAR), 89
    benchmark datasets
        discriminative-generative conflict, 95
        experiments, 95
        learning algorithm, 95
        strategy, 92
        trade-off, 92, 96, 97
        training and testing accuracy, 93
    binary classification, 90
    feature costs, 91
    gyroscope data, 90
    inference costs, 90
    pareto optimal configuration
        accuracy *vs.* inference costs, 93
        accuracy *vs.* sensor interfacing, 93

        inference cost, 93
        ScaleSI, 91
        test set, 92
        training set, 91, 92
    robustness, 92–93
    sensor costs, 90
    smartphone accelerometer, 90
Hybrid machine learning, 142


**I**
Inference, 10–11
Inference costs, 50
    PC, 50
Intelligent capabilities, 141
Internet-of-Things (IoT), 2
    cloud-centric deployment, 3
    definition, 2
    devices, 3
    edge layer, 3
    structure, 2


**J**
Joint probability distributions, 25


**L**
Logical models, 8–9
Logistic regression, 102
Low-quality feature extraction, 68


**M**
Machine learning, 139, 140
Machine learning pipeline
    accuracy, 6
    bringing hardware-awareness, 14
    features, 9–10
    fundamental trade-off, 14
    high-leve, 4
    models, 8–9
    overview, 4, 5
    pareto-optimal operating points, 15
    performance, 6
    tasks, 5
Mixed-signal sensor front-end
    resource *vs.* quality trade-offs, 44
Model complexity switching strategy
    activity recognition, 120
    Bayes-optimal conditional probability, 120
    conditional probability, 120
    level of difficulty, 120
    multi-class activity recognition task, 120
    policy, 120–121

selection, 121–122
switching policy
  Bayes-optimal conditional probability, 122
  conditional probability value, 125
  confusion matrices, 124
  estimated accuracy, 123
  human activity recognition, 123
  simple-to-complex parameter, 125
  thresholds, 125
  training dataset, 123
time aspects
  prediction and policy frequency, 126, 127
  prediction stability, 127–128

**N**
Naive Bayes (NB), 102
Noise-scalable Bayesian network (ns-BNs), 140
  assumptions, 77
  classification accuracy, 58
  device configurations, 119
  digital overhead costs, 118–119
  expressiveness and general discriminative abilities, 120
  five-feature, 56, 57
  hardware-aware learning, 119
  inference, 58, 119
  model, 56–57
  Pareto-optimal selection
    Accelerometer-X, 113
    Bayesian network, 116
    dynamic conditions, 114
    Gyroscope-X off, 113
    human activity recognition benchmark, 113
    Look-Up-Table (LUT), 113–114
    off-line implementation, 115
    optimal configuration, 114
    sampling process, 115
    ScaleFeatureNoiseRT algorithm, 114, 115
  probabilistic models, 113
  remaining challenges, 119
  robustness to missing features, 116–118
  scalable sensor front-end, 112
  search strategy, Pareto-optimal front (*see* Pareto-optimal trade-off)
  sensor malfunction, 112
  six-class HAR classification, 89, 145
  synthetic dataset for digital scaling ns-BN, 145

TAN classifier, 57
use cases
  adaptive sensing approaches, 75–76
  analog quality scaling, 71–75
  digital quality scaling, 68–71 (*see also* Digital scaling ns-BN)
  mixed-signal quality scaling, 62–67
  scalable blocks, 61
  sensor front-ends, 61
  structure and parametrization, 61
  ultra-low power VAD, 61
versatility, 62, 77

**O**
Operation-cost definitions, 51

**P**
Pareto-optimal front, 17, 18
Pareto-optimal trade-off
  extraction
    configuration selection, 88
    dataset pre-processing, 89
    feature and sensor set scaling, 86
    HAR (*see* Human activity recognition (HAR))
    inference costs, 89
    model complexity scaling, 86
    model learning, 89
    precision scaling, 86
    PrunePC, 88
    ScaleSI, 87
    search strategy, 87–88
    system configuration, 86
    system properties, 86
  noise configuration, 60
  objective function (OF), 59–60
  roles, ns-BN, 58
Practical feasibility, 141
Precision reduction
  hyperparameters, 153–158
Precision scaling, 129
Probabilistic circuits (PC), 23
  classification tasks, 37
  computational semantics, 35
  framework, 35
  properties, 35
  structural constraints
    decomposability, 36
    determinism, 37
    smoothness, 36–37
  structure and parameters, 35

Probabilistic inference, 26–27
Probabilistic models, 8, 15–16, 139
Probabilistic Sentential Decision Diagrams
        (PSDDs), 82
    Bayesian classifiers, 101
    Boolean functions, 83
    class conditional constraint, 95
    context specific independence, 84
    decision node, 83
    discriminative bias
        binary classification task, 98
        class variable, 97
        conditional probability query, 96
        D-LearnPSDD algorithm, 98, 99
        joint probability distributions, 97
        partial assignments, 98
        probability distribution, 99
        propositional variables, 98
        root decision node, 98
        root-node, 97
        vtree, 99, 100
    discriminative classifier, 101
    discriminative performance, tree
        learning approaches, 103
        learning strategy, 103
        log-likelihood and accuracy, 103
        probability distribution, 103
    D-LearnPSDD, 95, 102, 104
    experimental setup, 101
    generative bias, 101
    LearnPSDD algorithm, 84, 98
    naive Bayes (NB), 95
    polynomial time, 84
    probability distribution, 83
    random variables, 82
    robustness to missing features, 105
    structural constraints, 83
Probability theory
    distribution, 25
    joint probability distributions, 25
    notions and notation, 24–25
    statistical data, 24
Problem statement, 14–15


R
Regression, 6
Resource-constrained extreme-edge devices,
        139
Resource-constrained system, 141
Resource-efficient machine learning
    feature extraction stage, 11
    implementations, 12

inference, 10–11
    mobile activity recognition, 12
    run-time scenarios, 13
    sensor cost, 13
Resources and workloads, 141
Run-time strategies
    accelerometer, 132, 134
    cost-sensitive learning, 135
    dataset pre-processing, 128–129
    device configurations, 111
    dynamic modeling, 136
    dynamic run-time changes, 135
    hardware-aware probabilistic models,
        111
    human activity recognition benchmark,
        112
    hybrid architecture, 136
    hyperparameters, 130–131
    implications, 111
    machine learning algorithms, 134
    malfunctioning sensors, 111
    model-complexity tuning strategy, 112
    model learning, 129
    model selection, 121–122
    model switching strategy, 132
    neural networks, 135
    Noise-scalable Bayesian network (*see*
        Noise-scalable Bayesian network)
    occlusive environmental conditions, 111
    off-line optimization, 111
    probabilistic models, 111, 136
    real-time extreme-edge applications, 135
    selective execution, 135
    signal quality, 111
    state-of-the-art classifiers, 135
    state-of-the-art models, 136
    trade-off search strategy, 112
    tuning model complexity, 136
Run-time tuning, 51


S
Scalable hardware, 42, 55, 136
Scalable mixed-signal noise, 61
Scalable sensor front-end
    ADC and digital feature extraction block,
        61
    analog quality scaling, 71–75
    digital quality scaling, 68–71
    hardware design choices, 55
    mixed-signal quality scaling, 62–67
    with tunable amplification circuitry, 61
    ultra-low power VAD, 61

Semiconductor technology, 139
Sensing cost
    definition, 43
Sensor streams, 81
Sensory embedded pipeline
    building blocks, 38
    dynamic tuning block, 39
    energy consumption, 38
    hardware-aware probabilistic model, 39
    tunable feature extracting front-end, 39
Sentential Decision Diagrams (SDDs), 83
Stable, 128
State-of-the-art, 139
Supervised tasks, 6
System-wide hardware-awareness, 16

T
Time-series aspects, 142
Tractable probabilistic models, 34
Tree Augmented Naive Bayes (TANB), 34, 95
    classifiers, 34
Tunable hardware-induced noise, 58

U
Unsupervised tasks, 6

V
Voice activity detection (VAD), 61, 72–75

Printed in the United States
by Baker & Taylor Publisher Services